Guidance Note 1

Selection & Erection

17th IET Wiring Regulations Seventeenth Edition BS 7671:2008 Requirements for Electrical Installations incorporating Amendment Number 3:2015, referenced in the text as BS 7671:2008 + A3:2015

Published by The Institution of Engineering and Technology, London, United Kingdom

The Institution of Engineering and Technology is registered as a Charity in England & Wales (no. 211014) and Scotland (no. SC038698).

The Institution of Engineering and Technology is the institution formed by the joining together of the IEE (The Institution of Electrical Engineers) and the IIE (The Institution of Incorporated Engineers). The new Institution is the inheritor of the IEE brand and all its products and services, such as this one, which we hope you will find useful.

First published 1992 (0 85296 535 4)

Reprinted (with amendments) 1993

Second edition (incorporating Amendment No. 1 to BS 7671:1992) 1996 (0 85296 864 7)

Third edition (incorporating Amendment No. 2 to BS 7671:1992) 1999 (0 85296 954 6)

Fourth edition (incorporating Amendment No. 1 to BS 7671:2001) 2001 (0 85296 989 9)

Reprinted (incorporating Amendment No. 2 to BS 7671:2001) 2004

Fifth edition (incorporating BS 7671:2008) 2009 (978-0-86341-855-6)

Sixth edition (incorporating Amendment No. 1 to BS 7671:2008) 2012 (978-1-84919-271-2)

Reprinted 2012, 2014

Seventh edition (incorporating Amendment Nos. 2 and 3 to BS 7671:2008) 2015 (978-1-84919-869-1)

Copies of this publication may be obtained from:
The Institution of Engineering and Technology
PO Box 96, Stevenage, SG1 2SD, UK
Tel: +44 (0)1438 767328
Email: sales@theiet.org
www.electrical.theiet.org/books

ISBN 978-1-84919-869-1 (paperback)
ISBN 978-1-84919-870-7 (electronic)

Typeset in the UK by the Institution of Engineering and Technology, Stevenage
Printed in the UK by Sterling Press Ltd, Kettering

Contents

Chapter 5   Installation of cables   75

Chapter 6   Sizing of cables   89

Chapter 7   Other influences   93

Chapter 8   Installation of equipment   113

Cooperating organisations

The Institution of Engineering and Technology acknowledges the invaluable contribution made by the following individuals in the preparation of this Guidance Note:

Eur Ing K. Armstrong CEng FIEE ACGI
M. Coles BEng(Hons) MIET
P.R.L. Cook CEng FIEE
G.D. Cronshaw CEng FIET
P.E. Donnachie BSc CEng FIET
P.E. Donnachie BSc CEng FIET
Eur Ing G. Kenyon CEng MIET
H.R. Lovegrove IEng FIET
Eur Ing L. Markwell MSc BSc(Hons) CEng MIEE MCIBSE LCGI
T.R. Pickard IEng MIET
I.M. Reeve BTech CEng MIEE
R. Townsend CMS MIET

We would like to thank the following organisations for their continued support:

BEAMA Installation Ltd
BEAMA Ltd
Certsure trading as NICEIC and Elecsa
Electrical Contractors' Association
Electrical Contractors' Association of Scotland (SELECT)
Health and Safety Executive
IET - members of the User's Forum
NAPIT

Guidance Note 1 revised, compiled and edited

J. Bradley BSc CEng FIET FCIBSE

Acknowledgements

References to British Standards, CENELEC Harmonization Documents and International Electrotechnical Commission (IEC) standards are made with the kind permission of BSI. Complete copies can be obtained by post from:

BSI Customer Services
389 Chiswick High Road, London W4 4AL
Tel: +44 (0)20 8996 9001
Fax: +44 (0)20 8996 7001
Email: orders@bsi-global.com

The BSI also maintains stocks of international and foreign standards, with many English translations. Up-to-date information on BSI standards can be obtained from the BSI website: www.bsi-global.com.

Illustrations of test instruments were provided by Rod Farquhar Design: www.rodfarquhar.co.uk

Cover design and illustration were created by The Page Design: www.thepagedesign.co.uk

It is strongly recommended that anyone involved in work on or near electrical installations possesses a copy of the *Memorandum of guidance on the Electricity at Work Regulations 1989* (HSR25) published by the Health and Safety Executive. Copies of Health and Safety Executive documents and approved codes of practice (ACOP) can be obtained from:

HSE Books
PO Box 1999, Sudbury, Suffolk CO10 2WA
Tel: +44 (0)1787 881165
Email: hsebooks@prolog.uk.com
Web: http://books.hse.gov.uk

RCD terminology and information was provided by the Low Voltage Circuit Breaker Division of BEAMA and is taken from the BEAMA *Code of Practice for the Selection and Application of Residual Current Devices*.

Information on the forms of separation of LV switchgear were provided by the Low Voltage Distribution Switchboard Division of BEAMA. The BEAMA booklet *Guide to forms of separation* can be obtained from BEAMA, Westminster Tower, 3 Albert Embankment, London SE1 7SL. Tel: 020 7793 3000.

Extracts from the LDSA document *Fire safety guide No. 1* on Section 20 buildings are reproduced with permission from The London District Surveyors Association: www.londonbuildingcontrol.org.uk.

The tables on colour identification of buried services are reproduced with the permission of the National Joint Utilities Group: www.njug.org.uk.

The tables on cable references in Appendix E were reproduced with permission of Anixter (UK) Ltd, 1 York Road, Uxbridge, Middlesex UB8 1RN.

Preface

This Guidance Note is part of a series issued by the Institution of Engineering and Technology to explain and enlarge upon the requirements in BS 7671:2008, the 17th Edition of the IET Wiring Regulations, incorporating Amendments No. 1: 2011, No. 2: 2013 and No. 3: 2015.

Note that this Guidance Note does not ensure compliance with BS 7671. It is intended to explain some of the requirements of BS 7671, but readers should always consult BS 7671 to satisfy themselves of compliance.

The scope generally follows that of BS 7671; the relevant Regulations and Appendices are noted in the margin. Some Guidance Notes also contain material not included in BS 7671:2008+A3: 2015 but which was included in earlier editions of the Wiring Regulations. All of the Guidance Notes contain references to other relevant sources of information.

Electrical installations in the United Kingdom that comply with BS 7671 are likely to satisfy Statutory Regulations such as the Electricity at Work Regulations 1989. However, this cannot be guaranteed. It is stressed that it is essential to establish which Statutory and other Regulations apply and to install accordingly. For example, an installation in premises subject to licensing may have requirements different from, or additional to, those of BS 7671 and these will take precedence.

Guidance Note 1: Selection & Erection
© The Institution of Engineering and Technology

Introduction

This Guidance Note is principally concerned with Part 5 of BS 7671 – Selection and erection of equipment.

Neither BS 7671 nor the Guidance Notes are design guides. It is essential to prepare a full design and specification prior to commencement or alteration of an electrical installation. Compliance with the relevant standards should be required.

514.9 The design and specification should set out the requirements and provide sufficient information to enable electrically skilled persons and electrically competent persons (where appropriate) to carry out the installation and to commission it. The specification must include a description of how the system is to operate and all the design and operational parameters. It must provide for all the commissioning procedures that will be required and for the provision of adequate information to the user. This will be by means of an operational manual or schedule and 'as fitted' drawings if necessary.

It must be noted that it is a matter of contract as to which person or organisation is responsible for the production of the parts of the design, specification, construction and verification of the installation and any operational information. The persons or organisations who might be concerned in the preparation of the works include:

The Designer
The Planning Supervisor
The Installer (or Contractor)
The Distributor of Electricity
The Installation Owner (Client) and/or User
The Architect
The Fire Prevention Officer
Any Regulatory Authority
Any Licensing Authority
The Health and Safety Executive.

132.1 In producing the design, advice should be sought from the installation owner and/or user as to the intended use. Often, as in a speculative building, the intended use is unknown. The specification and/or the operational manual must set out the basis of use for which the installation is suitable.

The operational manual must include a description of how the system as installed is to operate and all commissioning records. The manual should also include manufacturers' technical data for all items of electrical equipment, wiring, switchgear, accessories, etc. and any special instructions that may be needed.

The Health and Safety at Work etc. Act 1974 Section 6 and the Construction (Design and Management) Regulations 2007 are concerned with the provision of information.

Guidance on the preparation of technical manuals is given in BS EN 82079-1, *Preparation of instructions for use. Structuring, content and presentation. General principles and detailed requirements* and the BS 4940 series *Technical information on constructional products and services*. The size and complexity of the installation will dictate the nature and extent of the manual.

General requirements

1.1 General

Part 2 'Equipment' is short for 'electrical equipment' and is defined as:

> *Any item for such purposes as generation, conversion, transmission, distribution or utilisation of electrical energy, such as machines, transformers, apparatus, measuring instruments, protective devices, wiring systems, accessories, appliances and luminaires.*

120.2
133.5 Installation designers may need information from the manufacturer as to the suitability of equipment for its intended use. It is not the intention of BS 7671:2008 (2015) to
511.1 stifle innovation or new techniques but, generally, the standard only recognises and
511.2 considers established materials and techniques.

1.2 Equipment

120.3 The designer is responsible for the safety of the design. Any intended departures from BS 7671, although the designer is confident regarding safety, must be recorded on the Electrical Installation Certificate. The resulting degree of safety of the installation must not be less than that obtained by compliance with the Regulations.

Chap 13 Chapter 13 of BS 7671, 'Fundamental Principles', outlines the basic requirements. Later chapters describe in more detail particular means of compliance with Chapter 13. Chapter 13 is normally referred to only where it is intended to adopt a method not recognised in later chapters. The terms are chosen to allow for interpretation to suit special cases.

The phrase *so far as is reasonably practicable* is used in Parts 6 and 7 of the Regulations. It should be borne in mind that methods described in later chapters are considered reasonably practicable in most circumstances.

Where methods are used which are not described in later chapters, then the onus is on the designer to confirm that the degree of safety is not less than that required by Chapter 13.

1.3 The Electricity at Work Regulations 1989

The requirements of the Electricity at Work Regulations 1989 (EWR) are intended to provide for the safety of persons gaining access to or working with or near electrical equipment.

The Memorandum of guidance on the Electricity at Work Regulations 1989 (Health and Safety Executive publication HSR25) should be carefully studied and it should be borne in mind that the EWR apply to designers, installers and users of installations.

132.12 BS 7671 is intended for designers, installers and those verifying electrical installations.
Chap 34 The Regulations, therefore, include requirements for the design of the installation and the selection and erection of electrical equipment. The installation designer should assess the expected maintenance of the installation and the initial design should make provision for maintenance to be carried out. The user has the responsibility for ensuring that equipment is properly operated and maintained when necessary.

Regulation 4(2) of the EWR requires that adequate maintenance is undertaken to prevent danger and HSR25 advises that regular inspection of electrical systems (supplemented by testing as necessary) is an essential part of any preventive maintenance programme. Regular operational functional testing of safety circuits (emergency switching/stopping, etc.) may be required, as, unlike functional circuits, they may be used infrequently. Comprehensive records of all inspections and tests should be made, retained and reviewed for any trends that may arise.

GN3 Guidance Note 3 *Inspection & Testing* gives guidance on initial and periodic inspection and testing of installations.

1.4 The Construction (Design and Management) Regulations 2007

The Construction (Design and Management) Regulations 2007 (CDM Regulations) place responsibilities on most installation owners and their professional design teams to ensure a continuous consideration of health and safety requirements during the design and construction of, and throughout the life of, an installation, including maintenance, repair and demolition.

Almost everyone involved in construction work will have a legal duty placed on them under the regulations. Those with legal duties are commonly known as 'dutyholders'.

Dutyholders under the CDM Regulations are:

▶ Clients – A 'client' is anyone having construction or building work carried out as part of their business. This could be an individual, partnership or company and includes property developers or management companies for domestic properties.
▶ CDM coordinators – A 'CDM coordinator' has to be appointed to advise the client on projects that involve more than 30 days or 500 person days of construction work. The CDM coordinator's role is to advise the client on health and safety issues during the design and planning phases of construction work.
▶ Designers – The term 'designer' has a broad meaning and relates to the function performed, rather than the profession or job title. Designers are those who, as part of their work, prepare design drawings, specifications, bills of quantities and the specification of articles and substances. This could include architects, engineers and quantity surveyors.
▶ Principal contractors – A 'principal contractor' has to be appointed for projects which last more than 30 days or involve 500 or more person days of construction work. The principal contractor's role is to plan, manage and coordinate health and safety while construction work is being undertaken. The principal contractor is usually the main or managing contractor for the work.
▶ Contractors – A 'contractor' is a business who is involved in construction, alteration, maintenance or demolition work. This could involve building, civil engineering, mechanical, electrical, demolition and maintenance companies,

partnerships and the self-employed.
- ▶ Workers – A 'worker' is anyone who carries out work during the construction, alteration, maintenance or demolition of a building or structure. A worker could be, for example, an electrician as well as those supervising the work, such as chargehands and foremen.

1.5 Building Regulations

1.5.1 The Building Regulations 2000 (England and Wales)

A series of Approved Documents have been published for the purpose of providing practical guidance to the requirements of the Building Regulations 2010 (SI 2010 No. 2214) (as amended) for England and Wales. Approved Documents of particular interest to designers and installers of electrical installations are listed below. Different versions of the Approved Documents apply in England and Wales.

- ▶ A – Structure
- ▶ B – Fire safety, Volume 1: Dwellinghouses
- ▶ B – Fire safety, Volume 2: Buildings other than dwellinghouses
- ▶ C – Site preparation and resistance to contaminants and moisture
- ▶ E – Resistance to the passage of sound
- ▶ F – Ventilation
- ▶ L1A – Conservation of fuel and power (New dwellings)
- ▶ L1B – Conservation of fuel and power (Existing dwellings)
- ▶ L2A – Conservation of fuel and power (New buildings other than dwellings)
- ▶ L2B – Conservation of fuel and power (Existing buildings other than dwellings)
- ▶ M – Access to and use of buildings
- ▶ P – Electrical safety: Dwellings
- ▶ Regulation 7 – Materials and workmanship.

All of the Approved Documents can be downloaded from the Planning Portal at http://www.planningportal.gov.uk, which contains separate English and Welsh sites.

Of particular importance is Regulation 7 of the Building Regulations 2010, which deals with materials and workmanship and requires materials to be appropriate for their application and be properly applied in a workmanlike manner.

Approved Document B includes requirements for the maintenance of a fire barrier, fire stopping and protection of openings, and wall and ceiling linings (including thermoplastic lighting diffusers).

Approved Document E includes a requirement for the maintenance of acoustic resistance.

Approved Document F includes a requirement for maintenance of airtightness; see also BS 9250 *Code of practice for the airtightness of ceilings in pitched roofs*.

Approved Documents L1A and L1B contain a requirement that reasonable provision shall be made for occupiers to obtain the benefits of efficient lighting.

Approved Document M includes requirements for the mounting heights of accessories and these are referred to in Appendix C of this Guidance Note (see C7).

Lighting and electric heating

Determining the specification for lighting and, where applicable, electrical heating for compliance with the Building Regulations for England and Wales has become more of a balancing act than it used to be, particularly in view of the amendments of the Building Regulations introduced in 2013. The designer may take into account certain factors that contribute to a building's overall energy efficiency, such as relevant details relating to the building fabric.

Further information can be found in Approved Documents L and their respective Compliance Guides and also in the IET *Electrician's Guide to the Building Regulations*.

1.5.2 The Building (Scotland) Regulations 2004

The Building Regulations 2010 are not applicable in Scotland, where the Building (Scotland) Regulations 2004 (as amended) apply.

The detailed requirements are given in the Technical Standards for compliance with the Building (Scotland) Regulations.

Guidance on how to achieve compliance with the Standards set in the Regulations is given in two Scottish Building Standards Technical Handbooks – Domestic and non-domestic.

These handbooks contain recommendations for electrical installations including the following:

- ▶ Compliance with BS 7671
- ▶ Minimum number of socket-outlets in dwellings
- ▶ Minimum number of lighting points in dwellings
- ▶ Minimum illumination levels in common areas of domestic buildings, for example, blocks of flats
- ▶ Minimum mounting heights of switches and socket-outlets, etc.
- ▶ Separate switching for concealed socket-outlets, for example, behind white goods in kitchens
- ▶ Conservation of fuel and power in both domestic and non-domestic buildings.

With regard to electrical installations in Scotland, the relevant requirements of the above Building Regulations are deemed to be satisfied by complying with BS 7671.

The handbooks may be downloaded from the Scottish Building Standards Division (BSD) website: www.scotland.gov.uk/bsd.

The IET Electrician's Guide to the Building Regulations also includes guidance relating to the Scottish Building Regulations (in addition to that relating England and Wales).

1.6 Competence

301.1
Appx 5
Part 2

BS 7671 requires that an assessment be made of the external influences to which the installation is to be exposed.

Amongst these, as indicated in Appendix 5 of BS 7671, is Category BA – classification of the capability of persons. In terms of electrical competence, Category BA includes three different levels: Skilled Person, Instructed Person and Ordinary Person. The definitions for these, as given in Part 2 of BS 7671, are:

Skilled person (electrically). Person who possesses, as appropriate to the nature of the electrical work to be undertaken, adequate education, training or practical skills, and who is able to perceive risks and avoid hazards which electricity may can create.

Instructed person (electrically). Person adequately advised or supervised by a skilled person (as defined) to enable that person to perceive risks and to avoid hazards which electricity can create.

Ordinary person. Person who is neither a skilled person nor an instructed person.

NOTE: The term "(electrically)" is assumed to be present where the terms 'skilled person' or 'instructed person' are used throughout BS 7671.

EWR Regulation 16   To require, or allow, a person who is not competent to undertake electrical work may be a breach of statutory Health and Safety legislation, including the Electricity at Work Regulations 1989.

Regulation 16 of the Electricity at Work Regulations 1989 requires persons to be competent to prevent danger and injury. The HSE publication HSR 25 provides guidance on this.

Selection and erection of equipment

2.1 General

Part 5   Part 5 of BS 7671 gives the general requirements for the selection and erection of equipment.

Perhaps one of the most significant requirements is that contained in Section 511, Compliance with standards, discussed below.

2.2 Compliance with standards

511.1   BS 7671 recognises equipment which complies with a British or Harmonized Standard appropriate to the intended use of the equipment without further qualification.

Part 2   A Harmonized Standard is defined in Part 2 as:

> *A standard which has been drawn up by common agreement between national standards bodies notified to the European Commission by all member states and published under national procedures.*

There is a statutory definition of Harmonised Standard in the Supply of Machinery (Safety) Regulations 1992 (SI 1992 No. 3073). This states:

> *'Harmonised standard' means a technical specification adopted by the European Committee for Standardisation (CEN) or the European Committee for Electrotechnical Standardisation (CENELEC) or both, upon a mandate from the Commission in accordance with Council Directive 83/189/EEC of 28 March 1983 laying down a procedure for the provision of information in the field of technical standards and regulations, and of which the reference number is published in the Official Journal of the European Communities.*

2.2.1 European norms (ENs)

Appx 1   European Norm (EN) standards are standards that are required to be adopted by all CENELEC members and to be published with identical text by all members, without any additions, deletions or further technical amendments. Such EN standards then supersede the relevant national standards which are withdrawn to an agreed timescale. They are published as BS ENs in the UK.

2.2.2 Harmonization documents (HDs)

Harmonization Documents (HDs) forming the CENELEC 384 and, more recently, 60364 series are documents based on the IEC 60364 suite of standards.

Harmonized Documents are standards that have been agreed by all CENELEC member countries and National Committees are obliged to implement those standards.

A national standard based on an HD may have further technical additions, made by the national standards committee, but requirements are not to be deleted. Note that national standards are not to be in conflict with the HD.

Preface BS 7671 is based on a number of CENELEC HDs within the 384/60364 series (see the Preface to BS 7671) with extra specific technical material added.

133.1.1 As stated earlier, BS 7671 recognises equipment which complies with a British or
511.1 Harmonized Standard appropriate to the intended use of the equipment without further qualification.

Equipment to a foreign national standard may be used, but in these cases the foreign standard is required by Regulation 511.1 to be based on an IEC standard, and the designer or specifier must verify that the equipment is at least as safe as similar equipment complying with the relevant British or Harmonized Standard.

133.1.3 For equipment that is to be used which is not to a British or Harmonized Standard, the
511.2 designer of the electrical installation is required to take an engineering view on its suitability and it should provide the performance and degree of safety as required by BS 7671.

2.3   Operational conditions and external influences

Sect 133 Equipment of all types must be suitable for its situation and use.

Part 3 An assessment by the designer of the characteristics and conditions of the installation
Sect 512 will be necessary, including the requirements of Part 3 of BS 7671.

331.1 The installation must be designed to be suitable for all the relevant conditions and external influences foreseen, including electricity supply effects or effects on the supply.

Appx 5 Appendix 5 of BS 7671 details the system of classification of external influences
Chap 52 developed in IEC 60364-3 and the classification is indicated in parts of Chapter 52. While not in general use in the UK, the classification can serve as a helpful reminder of conditions to be considered.

331.1 Regulation 331.1 lists some examples of characteristics of equipment which may adversely affect the equipment to be installed, the electricity supply or other services.

This list is not exhaustive but shows some details of what the designer should consider. Equipment must operate safely; further, it should be efficient and correctly selected.

512.1.1 All equipment must be selected to accommodate the worst foreseeable conditions of
512.1.2 service that can be encountered even if such conditions happen rarely.
512.1.3

301.1 The assessment of general characteristics requires careful consideration of the purpose for which the installation is intended to be used. Further, the building, installation structure and the electricity supply characteristics, including voltage, current and frequency, must be noted during the first part of the design.

If, for example, there is doubt that a switch or circuit-breaker can be used with inductive or capacitive circuits, e.g. motors, transformers or fluorescent lighting, advice should be obtained from the manufacturer.

2.3.1 Motors

512.1.4 Electric motors may have similar power ratings but differing applications. Motors for lifts, industrial plant and machinery, propulsion or ventilation will have differing duty cycles and will usually be the subject of a standard.

▼ **Figure 2.1** Induction motor

552.1.1 The fixed wiring must be able to match the duty cycle of the connected load. Infrequently used motor-driven equipment with brakes will have different demands on the fixed wiring when compared with, for example, a hydro-extractor that uses 'plugging' as a method of braking the equipment frequently. The demands on the fixed wiring should be established from the manufacturer's installation instructions. These details will form part of the installation manual. (See Guidance Note 6 for detailed information on cable selection for motor circuits.)

2.3.2 Electromagnetic compatibility (EMC)

515.1 All equipment must be selected and erected to permit safe working, prevent harmful effects to other equipment and not impair the supply arrangements. This includes the consideration of electromagnetic compatibility effects, as well as more straightforward considerations such as loading, voltage and current rating, and circuit arrangements. Amendment No. 1 to BS 7671:2008 introduced new Section 444, Measures against electromagnetic disturbances, and this is discussed in Chapter 7 of this Guidance Note. See also section 2.5.

2.3.3 Harmonics

Harmonics are an aspect of EMC. Harmonic voltages and currents can cause interference with the normal operation of equipment and overload cables in certain cases (see sections 2.5 and 6.3).

2.4 Identification and notices

2.4.1 Cables and conductors

Table 51 The means of identification for cables and conductors for fixed wiring are given in Table
514.3.1 51 of BS 7671 – reproduced here as Table 2.1 for convenience. Identification may be by colour and/or alphanumeric marking.

514.3.2 Cores of cables should be correctly identified at their terminations, and preferably throughout their length.

514.4.2 The colour combination green-and-yellow is reserved exclusively for identification of protective conductors and must not be used for any other purpose. Bare conductors and busbars which are to be used as protective conductors may be identified using green-and-yellow sleeving or tape, as appropriate.

▼ **Table 2.1** Identification of cables and conductors (from Table 51 of BS 7671)

Function	Alphanumeric	Colour
Protective conductors		Green-and-yellow
Functional earthing conductor		Cream
a.c. power circuit[1]		
Line of single-phase circuit	L	Brown
Neutral of single- or three-phase circuit	N	Blue
Line 1 of three-phase a.c. circuit	L1	Brown[5]
Line 2 of three-phase a.c. circuit	L2	Black[5]
Line 3 of three-phase a.c. circuit	L3	Grey[5]
Two-wire unearthed d.c. power circuit		
Positive of two-wire circuit	L+	Brown
Negative of two-wire circuit	L-	Grey
Two-wire earthed d.c. power circuit		
Positive (of negative earthed) circuit	L+	Brown
Negative (of negative earthed) circuit[2]	M	Blue
Positive (of positive earthed) circuit[2]	M	Blue
Negative (of positive earthed) circuit	L-	Grey
Three-wire d.c. power circuit		
Outer positive of two-wire circuit derived from three-wire system	L+	Brown
Outer negative of two-wire circuit derived from three-wire system	L-	Grey
Positive of three-wire circuit	L+	Brown
Mid-wire of three-wire circuit[2,3]	M	Blue
Negative of three-wire circuit	L-	Grey
Control circuits, ELV and other applications		
Line conductor	L	Brown, Black, Red, Orange, Yellow, Violet, Grey, White, Pink or Turquoise
Neutral or mid-wire[4]	N or M	Blue

Notes to Table 2.1:

1 Power circuits include lighting circuits.
2 M identifies either the mid-wire of a three-wire d.c. circuit, or the earthed conductor of a two-wire earthed d.c. circuit.
3 Only the middle wire of three-wire circuits may be earthed.
4 An earthed PELV conductor is blue.
5 Appendix 7 of BS 7671, item 5 states that in a two- or three-phase power circuit the line conductors may all be one of the permitted colours, either identified L1, L2, L3 or marked brown, black, grey at their terminations to show the line designation.

2.4.2   Circuits

514.9.1
621.1
A diagram, chart or table giving details of the circuits is required. The details must be used by the person verifying compliance with BS 7671; for example, details of the selection and characteristics of the protective devices are needed for verifying protection against overcurrent and electric shock. A durable copy of the details should be fixed in or adjacent to each distribution board. The same details will be needed to assist periodic inspection and testing at a later date.

631.1
632.1
632.2
For simple installations, the Electrical Installation Certificate, together with the Schedule of Inspections and the Schedule of Test Results, will meet the requirements provided each circuit is also identified at the distribution board.

2.4.3   Switchgear

514.1.1
514.8.1
Labelling of switchgear is very important, particularly where the route of the final circuit cables is not obvious. If there is any possibility of confusion, some reliable means of identification must be clearly visible. It is necessary for the protective devices to be marked so that they can be identified easily by the user (see BS EN 61439).

Remote operation of switchgear may be necessary and in this case the indicating equipment should be designed in accordance with BS EN 60447 and BS EN 60073 or equivalent.

514.1.1
621.1
Complex installations demand greater detail. Details of protective measures and cables should be provided as part of the 'as installed' information. When the occupancy of the premises changes, the new occupier should have sufficient information to correctly operate the electrical installation. Diagrams, charts, tables and schedules should be kept up to date. Such items are essential aids in the maintenance and periodic inspection and testing of an installation.

2.4.4   Warning notices

The warning notices called for in Section 514 of BS 7671 are intended to warn persons about the risk of working on or near live parts.

Voltage

Amendment No. 1 to BS 7671 has reaffirmed that for single-phase installations with a nominal voltage of 230 volts to earth and for three-phase installations with a nominal line-to-line voltage of 400 volts, no warning label is required.

514.10.1
In more unusual cases, where the nominal voltage (U_0) within an item of equipment or enclosure exceeds 230 volts (e.g. a non-standard voltage such as 650 V) and where

Notice, label or identification	Regulation
PV array and generator junction boxes labelled to warn that the equipment may still be live after isolation	712.537.2.2.5.1
Permitted characteristics and type of supply for mobile and transportable units	717.514
Caravan instructions for supply and periodic inspection	721.514.1
Notice inside caravan for connection to supply	721.537.2.1.1.1 Fig 721
Plan for heating system to be fixed at distribution board	753.514

2.5 Mutual detrimental influence

132.11
515.1
515.2
528.3

Regulation 515.1 requires that there be no harmful effect between electrical and other installations. The best approach, where practicable, is to arrange that the installations are kept separated. Damage can be caused by such influences as thermal effects, electrolysis or corrosion. The thermal effects of other installations, such as hot water systems, must be considered. Equipment should be either designed to operate properly at elevated temperatures, reduced temperatures, etc. or be protected from such effects. Electrolysis may result from leakage currents or from contact between dissimilar metals in damp conditions. Corrosion may result, for example, in rusting of unprotected steelwork; action needs to be taken to obviate these risks.

Sect 444

The Electromagnetic Compatibility (EMC) Regulations 2006 place a statutory requirement on designers and constructors to design/construct electrical equipment and systems so that they do not cause excessive electromagnetic disturbances (emissions) and are not unduly affected by electromagnetic interference from other electrical equipment or systems; see Chapter 7 of this Guidance Note.

Harmonic production and interference, electrostatic discharges, mains-borne signals, etc. are all types of interference that require consideration by the designer and installer.

2.6 Compatibility

133.4
332.1

All equipment should be selected and erected so that its intended function takes into account its susceptibility from other equipment. Further, the equipment must not cause emissions that impair the intended function of other equipment within or adjacent to the installation.

331.1

Regulation 331.1 is of particular importance when considering supplies to information technology equipment; this includes computers, electronic office equipment, data transmission equipment and point-of-sale financial transaction terminals. Switch-mode power supply units, used to power this type of equipment, are particularly well known for causing harmonic generation problems in addition to high protective conductor currents.

Whenever electrical equipment is switched on or off, particularly where inductive loads are involved, high-frequency voltage transients occur. This so-called mains-borne noise may cause malfunction in information technology equipment. The capacitance of the circuit cables and the filters incorporated in most information technology equipment attenuate this mains-borne noise. It is good practice to ensure that sources likely to give rise to significant noise, e.g. motors and thermostatically controlled equipment, are kept apart from sensitive equipment. Susceptible electronic equipment should be fed by separate circuits from the incoming supply point of the building. Additional filters (sometimes called 'power conditioners') may be used to reduce this transient noise on existing circuits and surge protective devices can be installed to divert or absorb transients (see section 3.9).

Large switch-on (inrush) current surges occur when transformers, motors, mains rectifier circuits, etc. are energized and can cause excessive short-time voltage drop in the circuit conductors (dips). Inrush currents affect other circuits also and may require larger cable sizes or the equipment to be put on separate circuits. It should be borne in mind that information technology equipment itself can cause the same problem if switched on in large groups.

Where maintenance of the supply voltage to especially sensitive equipment, such as information technology equipment, is considered of importance, the user may need a device such as a motor-generator or uninterruptible power supply (UPS).

For terminating cables in equipment the designer should use 70 °C cable unless he/she has confirmed that the equipment terminals are able to withstand greater temperatures.

2.7 Functional earthing

When planning installations consisting of information technology equipment, a dedicated conductor arrangement provided for functional earthing could be considered. Information on functional earthing is contained in chapter 7 and in Guidance Notes 5 and 8.

The protective conductors of the installation, which are connected to the main earthing terminal, are subject to voltages relative to the conductive mass of the Earth, of which two types of voltage appearing on the earthing system can be considered.

The first is transients. These are short-duration spikes on the voltage waveform which can be caused by load switching of high voltage (HV) equipment or lightning strikes. Transients may be generated by the charging of an equipment frame via the stray capacitance from the mains circuit or mains-borne transients may be coupled into the earthing conductor or frame from mains conductors.

The second type of voltage is created by switch-mode power supplies, which are designed to filter current to earth.

These voltages are termed earth noise, as any unwanted signal, or interfering voltage, is deemed to be 'noise'. Noise in the earthing system can cause disruption to sensitive microprocessors of computing equipment, as the earth-reference point is no longer 'zero'.

Manufacturers or designers of large computer systems usually make specific recommendations for the provision of a 'clean' mains supply or a 'clean' earth; these networks must always be connected to the building or installation earthing network.

A dedicated earthing conductor may be used for a computer system, provided that:

542.1.1 **(a)** all accessible exposed-conductive-parts of the computer system are earthed, the computer system being treated as an 'installation' where applicable

▼ **Figure 2.5**   Examples of Form 4 construction assemblies

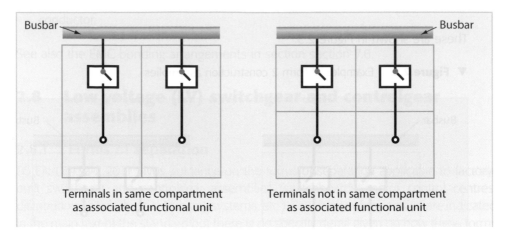

| Terminals in same compartment as associated functional unit | Terminals not in same compartment as associated functional unit |

2.8.2   General

Assemblies, including switchboards, panel boards and motor control centres, should be capable of withstanding the thermal and dynamic stresses resulting from fault currents.

The terminology to define the short-circuit rating of an assembly is given in the BS EN 61439 series of standards as follows:

▶ rated short-time withstand current, I_{cw}
▶ rated peak withstand current, I_{pk}
▶ rated conditional short-circuit current, I_{cc}

It is essential that the assembly manufacturer's ratings and instructions are followed.

IET Guidance Note 6 and the BEAMA Guide to verification both provide further guidance on this subject.

Tests, as detailed in BS EN 61439-1, are applied to the complete assembly as relevant at manufacture, including continuity, insulation resistance and perhaps a flash test at high voltage. The designer and installer must be aware that there are statutory requirements under the Electricity at Work Regulations and the CDM Regulations, etc., for the safe design, construction, operation and provision for maintenance of electrical equipment assemblies. Adequate access, working space and lighting need to be provided where work is to be carried out on or near equipment, in order that persons may work safely.

EWR 1989   It should also be appreciated that Regulation 8 of the Electricity at Work Regulations places an absolute requirement (one that shall be met regardless of cost or any other consideration) on protective conductor connections to earth:

> … *a conductor shall be regarded as earthed when it is connected to the general mass of earth by conductors of sufficient strength and current-carrying capability to discharge electrical energy to earth.*

It may be questioned whether the termination of steel or aluminium wire armour (where used as a protective conductor) with glands into metal gland plates, which themselves may only be bolted to the switchgear or controlgear frame, is an adequate connection. Cable gland 'earth tags' and supplementary connections to the equipment earth terminals may be necessary. In any event it should be ensured that any paint or other surface finish on the switchgear does not prevent effective electrical continuity between the adjacent parts.

612.13.2   During site installation and commissioning, as well as electrical testing, functional tests should be carried out on the complete assembly, plus any other specific tests advised by the manufacturer or required by the client, user or engineer. It is not usual to carry out a repeat of specialist tests, e.g. a flash test at site; however, in the event of such a requirement or request, the manufacturer's advice should be sought.

421.1.201   A particular requirement for consumer units and similar switchgear assemblies installed in domestic (household) premises was introduced in Amendment No 3 (2015) to BS 7671: 2011, by means of Regulation 421.1.201. The regulation requires such switchgear assemblies to comply with BS EN 61439-3 and to either: (i) have their enclosure manufactured from non-combustible material, or (ii) be enclosed in a cabinet or enclosure constructed of non-combustible material and complying with Regulation 132.12, relating to accessibility. Guidance Note 4 gives further information about the requirements.

▼ **Figure 3.2** Fusewire

3.5 Circuit-breakers

3.5.1 Circuit-breakers for applications not exceeding 440 V a.c.

Appx 3
411.4.5
There is a wide range of circuit-breaker characteristics that have been classified according to their instantaneous trip performance and Table 3.4 gives some information on the applications of the various types available. These limits are the maximum allowed in circuit-breakers to BS 3871 (now withdrawn) and circuit-breakers to BS EN 60898-1; note that manufacturers may provide closer limits. For such circuit-breakers, manufacturers' data for I_a may be applied to the formula given in Regulation 411.4.5. (I_a is the current causing automatic operation of the device within the stated time.)

▼ **Table 3.4** Circuit-breakers – overcurrent protection

Type	Multiple of rated current I_n below which it will not trip within 100 ms	Multiple of rated current I_n above which it will definitely trip within 100 ms	Typical application
1 B	2.7X 3X	4X 5X	Circuits not subject to inrush currents/switching surges
2 C	4X 5X	7X 10X	Circuits where some inrush current may occur. For general purpose use on fluorescent lighting circuits, small motors, etc.
C 3	5X 7X	10X 10X	Circuits where high inrush currents are likely, e.g. motors, large lighting loads, large air conditioning units
D 4	10X 10X	20X 50X	Circuits where inrush currents are particularly severe, e.g. welding machines, X-ray machines

Notes:
(a) Some non-linear resistive loads, such as large tungsten filament lighting installations, may give rise to high inrush currents.
(b) Due to the withdrawal of BS 3871, Types 1 to 4 circuit-breakers are now obsolete but are still to be found in use.
(c) Type D and Type 4 circuit-breakers are special-purpose circuit-breakers that require low earth fault loop impedance and should not be used without due consideration. Data should be obtained from the manufacturer.

3.5.2 Circuit-breakers for applications not exceeding 1000 V a.c. (MCCBs)

Unlike circuit-breakers to BS EN 60898-1, circuit-breakers complying with BS EN 60947-2 do not have defined characteristics and manufacturers' data must be used. BS EN 60947-2:2006 + A2: 2013 includes Annex L, which is for circuit-breakers not fulfilling the requirement for overcurrent protection (CBIs), derived from the equivalent circuit-breaker. A class X CBI is fitted with integral short-circuit protection, which may, on the basis of the manufacturer's data, be used in conjunction with the starter overload relay, for short-circuit protection.

3.5.3 Circuit-breakers – general

All circuit-breakers have a maximum fault current breaking capacity and care is needed in selection to ensure that this will not be exceeded in service. BS 3871 identified this capacity with an 'M' rating; however, in BS EN 60898 the 'M' category ratings for breaking capacities are replaced by an I_{cn} value embossed in a rectangle on the device, see Table 3.5. The manufacturer has to declare the breaking capacity of the devices at specified power factors of test current. Higher fault current capacities up to 25 kA are recognised for BS EN 60898 devices. Rated values for both standards are given in Table 3.5.

▼ **Table 3.5** Rated breaking capacity for miniature circuit-breakers to BS 3871, circuit-breakers to BS EN 60898 and RCBOs to BS EN 61009

Miniature CBs to BS 3871

'M' number marked on device	I_{cn} (A)
M1	1000
M1.5	1500
M3	3000
M4.5	4500
M6	6000
M9	9000

CBs to BS EN 60898 and RCBOs to BS EN 61009

Marked on device in rectangle	I_{cn} rated capacity (A)	k	I_{cs} service capacity (A)
1500	1500	1	1500
3000	3000	1	3000
4500	4500	1	4500
6000	6000	1	6000
10000	10000	0.75	7500
15000	15000	0.5	7500
20000	20000	0.5	10000
25000	25000	0.5	12500

Note: k is the ratio between service short-circuit capacity (I_{cs}) and rated short-circuit capacity (I_{cn}), BS EN 60898 and BS EN 61009.

Two short-circuit breaking capacities, I_{cn} and I_{cs}, are quoted for circuit-breakers:

I_{cn} the rated short-circuit capacity. This is the ultimate short-circuit breaking capacity.

I_{cs} the service short-circuit capacity. This is the maximum level of fault current operation after which further service is assumed without loss of performance.

For an assigned I_{cn} the I_{cs} value will not be less than the value tabulated in Table 3.5. The I_{cn} of the circuit-breaker must always exceed the prospective short-circuit

It may be appropriate to exclude certain circuits from additional protection by an RCD, an example of which is shown in Figure 3.5. This would not be permissible, however, where such protection is required by BS 7671 (see 3.1.4).

RCDs with a suitable rated residual operating current and operating time are suitable for:

411.1 ▶ fault protection

415.1 ▶ additional protection (in case of failure of basic protection or carelessness by users)

422.3.9 ▶ some protection against the risk of fire caused by fault currents.
705.422.7

RCDs are **intended to operate** to provide protection against line-to-earth and neutral-to-earth faults but **do not operate** in the event of line-to-line or line-to-neutral faults.

An RCD will only restrict the time during which a fault current flows. It cannot restrict the magnitude of the fault current, which depends solely on the circuit conditions.

411.4.4 For an installation which is part of a TN system the use of an RCD may be from choice
411.3.2.6 or because the earth fault loop impedance is too high for overcurrent protective devices to operate within the maximum permitted disconnection times. In the latter case, a further option is to use supplementary equipotential bonding so that there will be no dangerous potential differences between simultaneously accessible exposed- and extraneous-conductive-parts.

3.7.4 Additional protection

415.1 As indicated in 3.1.4, the RCD must have a rated residual operating current not exceeding 30 mA and an operating time not exceeding 40 ms at a residual current of $5 I_{\Delta n}$. This requirement is met by general type devices complying with BS EN 61008 or BS EN 61009 (or BS 4293).

411.4.9 A popular misconception about the additional RCD protection specified by Regulation 411.3.3 is that where RCDs or RCBOs are used, the circuit characteristics for disconnection times in Table 41.1 are required to be satisfied by both an overcurrent device and the RCD. This is not the case as highlighted in Regulation 411.4.9: the overcurrent device (or the overcurrent characteristic of an RCBO) may be used satisfy the overcurrent protection requirements only, with the RCD (or the RCD characteristic of the RCBO) being used to satisfy the requirements for automatic disconnection in the event of a fault, in accordance with Table 41.5.

3.7.5 Location of RCDs

RCDs must be part of the fixed installation and can often be located in consumer units or distribution boards in the form of RCCBs or RCBOs.

3.7.6 Selection of types of RCD

314.1 It is not good practice to provide a single RCD on the supply to a complete distribution
531.4.1 board/consumer unit, which may have lighting circuits, etc., as operation of the RCD would interrupt the supply to all circuits supplied by that distribution board/consumer unit. Regulation 314.1(i) requires installation circuit arrangements to be such that dangers are avoided and inconvenience minimized in the event of a fault. It can be argued that protection of an installation by an RCD is not related to the 'circuit' definition in Part 2 of BS 7671 and Regulation 531.4.1 makes specific reference to a single RCD

for a TT installation. However, the decision to provide a single RCD, to use a split distribution board/consumer unit or to use RCBOs must be taken only after due consideration of all the relevant factors, including convenience and maintenance, and not just initial costs. For example, elderly or infirm persons may not be able to gain access to, or reset, an RCD. A single 30 mA RCD should normally not be installed at the origin of the installation.

3.7.7   TT systems

411.5.2   BS 7671 recognises that, in a TT system, either RCDs or overcurrent protective devices
411.5.3   may be used to provide fault protection by automatic disconnection of supply. In many such installations it will not be practicable to attain a sufficiently low value of earth electrode resistance to use overcurrent protective devices.

▼ **Figure 3.6**   Installation forming part of a TT system

411.5.3   Where an RCD is used (see Figure 3.6), the product of the rated residual operating
Table 41.5   current ($I_{\Delta n}$), in amperes, and the sum of the earth electrode and protective conductor resistances (R_A), in ohms, must not exceed 50 V. This does not mean, however, that the voltage to earth will not exceed the prescribed voltage the RCD will operate. In any event, to ensure compliance with the requirements of Regulation 411.5.3 the earth fault loop impedance must not exceed the appropriate limit given in Table 41.5 of BS 7671.

3.7.8   Two distinct categories

From the foregoing, it is seen that it is possible to categorise RCDs into two distinct categories according to their fault current operating characteristics:

422.3.9   **(a)** RCDs having a rated residual operating current greater than 30 mA. This group
705.422.7   is intended solely to provide fault protection or protection against fire. 100 mA and 300 mA RCDs are sometimes said to protect against fire in the event of

3.9.6.4 Critical length of connecting conductors for SPDs

534.2.9 To gain maximum protection the connecting conductors to SPDs must be kept as short as possible, to minimize additive inductive voltage drops across the conductors. For SPDs installed in shunt or in parallel with the supply (see Figure 3.12), the total lead length $(a + b)$ should preferably not exceed 0.5 m, but should in no case exceed 1.0 m.

If long lead lengths cannot be avoided, it may be necessary to select an SPD with a lower voltage protection level, in order to provide efficient protection.

Where an SPD is fitted in-line (see Figure 3.13), the protective conductor (length c) should not exceed 0.5 m in length, and in no case should exceed 1.0 m. Refer to the SPD manufacturer's instructions for optimal installation.

Fig 53.3 ▼ **Figure 3.12** Critical length of connecting conductors for shunt installed SPDs

OCPD = overcurrent protective device
SPD = surge protective device
E/I = equipment or installation to be protected against overvoltages

Fig 53.4 ▼ **Figure 3.13** Critical length of connecting conductor for series or in-line installed SPDs

OCPD = overcurrent protective device
SPD = surge protective device
E/I = equipment or installation to be protected against overvoltages
NOTE: The line terminal on the SPD must be capable of carrying the load current.

External influences 4

4.1 General

301.1(ii) The effect of environmental conditions and general characteristics around various parts of the installation must be assessed to enable suitable electrical equipment to be specified.

Chap 51 A list of external influences relating to Chapter 32 of BS 7671 is given in Chapter 51
Appx 5 and Appendix 5. Many of these external influences are also identified in Section 522.
Sect 522 Chapter 52 of BS 7671 applies to the selection and erection of wiring systems but the classification of external influences is applicable and relevant to all types of electrical equipment.

134.1.1 If every part of the wiring system complies with the IP degree of protection for the worst circumstances expected in the particular location and correct measures are taken during installation and maintenance, the selection and erection requirements of BS 7671 with respect to external influences will generally be satisfied. Due consideration must be given to the installation of equipment, including fixings, and manufacturers' installation instructions complied with, or the declared IP rating of the equipment may not be achieved. Good workmanship is, as always, required.

512.2.1 Equipment must be of a design appropriate to the situation in which it is to be used or its mode of installation must take account of the conditions likely to be encountered.

All electrical equipment selected must be suitable for its location of use and method of installation and equipment should not be modified on site unless it is designed and manufactured to allow this. Manufacturers' recommendations and instructions should be followed. BS 7540-2:2005 *Electric cables. Guide to use for cables with a rated voltage not exceeding 450/750 V* gives general cable selection guidance.

Equipment must also be protected against damage and ingress of foreign bodies during construction work. This damage and ingress may be more severe than operational conditions.

References are made in the following sections to certain aspects covered in Section 522. The coding used in Appendix 5 of BS 7671 is given in parentheses.

4.2 Ambient temperature (AA)

Part 2 Local ambient temperature means the temperature of the air or other medium around the equipment.

522.1 There can be a wide range between the highest and lowest temperatures around equipment, depending on time of year, artificial heating, etc. A design must be based on the most extreme ambient envisaged. In the UK, external ambient temperatures between −10 °C and +35 °C commonly occur and occasionally even lower or higher temperatures than these.

Appx 4
Table 4B1
Table 4B2
The cable current-carrying capacity tables in Appendix 4 of BS 7671 are based on an ambient air temperature of 30 °C and, for buried cables, an ambient ground temperature of 20 °C. Where these temperatures differ, rating factors are to be used. Appendix 4, clause 2.1 and Tables 4B1 and 4B2 should be consulted.

While the air temperature in most installations in the UK would not usually be over 30 °C, close to a domestic heating appliance or in an industrial environment it may be higher. BS 1363 accessories and cable with an operating temperature rating of 70 °C or lower are not recommended for installation in areas of high ambient temperature and in particular must not be covered with clothes or other items. Heat-resisting cables appropriate to the ambient and equipment temperatures must be utilized.

753.522.1.3 Cables in heated floors or other parts must be suitable for the highest temperature reached.

Cables and equipment must also be suitable for the lowest likely temperature. In particular, general-purpose thermoplastic cables (e.g. PVC) should not be installed in cold places. Attention is drawn to the fact that, as the temperature decreases, thermoplastic compounds become increasingly stiff and brittle, with the result that, if the cable is bent quickly to a small radius or is struck at temperatures in the region of 0 °C or lower, there is a risk of shattering the thermoplastic components.

To avoid the risk of damage during handling, it is desirable that thermoplastic insulated or sheathed cables should be installed only when both the cable and the ambient temperature are above 0 °C and have been so for the previous 24 hours, or where special precautions have been taken to maintain the cables above this temperature. BS 7540 advises a minimum ambient temperature of 5 °C for some types of thermoplastic insulated and sheathed cables.

4.3   Solar radiation (AN) and ultraviolet radiation

522.2.1
522.11.1
Where cables are installed in situations subject to solar gain or other radiant heat sources, they should be resistant to damage from that source or effectively protected from damage. Cables may need to be derated where there are heat sources or heat is absorbed by solar gain.

To protect cables from the adverse effects of ultraviolet (UV) radiation, the sheaths often contain carbon black or alternative UV stabilisers. Physically shielding the cables from direct solar radiation may also be necessary. In situations where high ambient temperatures are present (e.g. hot countries) cables will need to be derated. Shielding may be necessary, to protect from both direct heat and heat absorbed from solar radiation but cables should be freely ventilated. Cable manufacturers will provide guidance on the use of cables outdoors.

Where weather resistance is required, the type of cable sheath to give protection should be carefully considered. Cable sheathing materials should conform to BS 7655, parts of which have been replaced by parts of BS EN 50363. Protection from mechanical damage may also be required. Further advice is given in BS 7540, parts of which have been replaced by parts of BS EN 50565.

The exposure of thermoplastic insulation or sheathing to high temperatures for any length of time will lead to softening. If such a temperature is maintained for long periods, the thermoplastic can decompose and give off corrosive products which may attack the conductors and other metalwork. Softening of the thermoplastic may also allow the conductor to move through the insulation if any mechanical force is applied (e.g. cables hanging vertically). Decentralisation of conductors may also take place where cables are overheated when bunched inside trunking.

4.4 The IP and IK codes

512.2 The designer must assess each area of external influence then ensure the required classifications are selected and applied correctly.

The IP code, BS EN 60529, describes a system for classifying the degrees of protection provided by the enclosures of electrical equipment. Superseded descriptions such as 'ordinary', 'drip-proof', 'splashproof' and 'watertight', are not defined terms.

The IK codeBS EN 62262, describes a system for classifying the degrees of protection provided by enclosures for electrical equipment against external mechanical impacts. The letters IK are followed by two numerals which identify a specific impact energy.

Appendix B of this Guidance Note lists the degrees of protection indicated by these codes.

4.5 Presence of water (AD) or high humidity (AB)

522.3 Any wiring system or equipment selected and installed must be suitable for its location and able to operate satisfactorily without undue deterioration during its working life. The presence of water can occur in several ways, e.g. rain, splashing, steam/humidity, condensation, and at each location where it is expected to be present its effects must be considered. Suitable protection must be provided, both during construction and for the completed installation.

522.5 In damp situations and wherever they are exposed to the weather, all metal sheaths and armour of cables, metal conduit, ducts, trunking, clips and their fixings should be of corrosion-resistant material or finish. They should not be in contact with other metals with which they are liable to set up electrolytic action (see section 4.7 also).

In any situation and during installation, the exposed insulation at terminations and joints of cables which are insulated with hygroscopic material, e.g. mineral insulated cable, must be sealed against the ingress of moisture. Such sealing material and any material used to insulate the conductors where they emerge from the cable insulation should have adequate insulating and moisture-proofing properties and retain those properties throughout the range of service temperatures.

The manufacturer's recommendations regarding the termination and installation of all types of cable must be strictly observed.

Sect 526 In a damp situation, enclosures for cores of sheathed cables from which the sheath has been removed and for non-sheathed cables at terminations of conduit, duct, ducting or trunking systems should be damp-proof and corrosion-resistant. Every joint in a cable should be suitably protected against the effects of moisture.

522.3.2 Conduit systems not designed to be sealed should be provided with drainage outlets at any points in the installation where water might otherwise collect. Those outlets must not affect the electrical safety of the system, or allow vermin access.

512.2.1 If it is necessary to locate a switch, socket-outlet, etc. near to a sink, the device must be protected from any water splashing that may occur. This can be achieved either by locating the accessory in a position where it will not be exposed to splashing, or providing a splash-resistant covering, or providing a weather protected device.

(Further information is contained in Appendix C6.2 'Sinks and electrical accessories'.)

It may occasionally be necessary to consider the installation of cables in locations that permanently contain water, such as fountains and marinas. It must be noted that plastic insulating and sheathing materials of cables are not totally waterproof and will absorb some water over time. Such cables are not suitable for continuous immersion and the cable manufacturer's advice should be taken for such installations. (See Guidance Note 7 for further information.)

4.6 Presence of solid foreign bodies (AE)

522.4 Where dusty conditions may be expected to occur, equipment and enclosures for conductors and their joints and terminations should have the degree of protection of at least IP5X. There are two conditions recognised in this particular IP code, the first condition being where the normal working cycle of the equipment concerned causes reductions in air pressure, e.g. thermal cycling effects, and the second condition being where such pressure reductions do not occur. These conditions should be considered during the design and selection procedure and appropriate equipment chosen.

It is necessary to have good operational housekeeping policies to ensure dust is kept to a minimum. A build-up of dust on electrical equipment, cables, etc., can act as thermal insulation and cause overheating of the equipment or cable. The dust may even be of a type that can ignite if the temperature of equipment rises. However, combustible dusts are outside the scope of this Guidance Note and reference should be made to the BS EN 50281 series, the BS EN 61241 series or the BS EN 60079 series, as applicable, for further guidance.

To assist in cleaning in dusty installations, a design should be developed in which inaccessible surfaces and locations are minimized and surfaces that can collect dust are made as small as practicable, or sloped to shed dust. Cable ladder rack is preferred to cable tray and ladder rack should be suspended vertically (i.e. on edge) rather than horizontally (i.e. flat).

4.7 Presence of corrosive or polluting substances (AF)

522.5 In damp situations, where metal cable sheaths and armour of cables, metal conduit and conduit fittings, metal ducting and trunking systems and associated metal fixings are liable to chemical or electrolytic attack by materials of a structure with which they may come into contact, it is necessary to take suitable precautions against corrosion, such as galvanizing or plating.

Materials likely to cause such attack include:

▶ materials containing magnesium chloride which are used in the construction of floors and plaster mouldings
▶ plaster coats containing corrosive salts

- lime, cement and plaster, for example on unpainted walls, or over cables buried in chases
- oak and other acidic woods
- dissimilar metals liable to set up electrolytic action.

Application of suitable coatings before erection, or prevention of contact by separation with plastics, are recognised as effective precautions against corrosion.

Contact between bare aluminium conductors, aluminium sheaths or aluminium conduits and any parts made of brass or other metal having a high copper content should be avoided, especially in damp situations, unless the parts are suitably plated. If such contact is unavoidable, the joint should be completely protected against ingress of moisture. Modern, friction welded bimetallic joints are available for most tape-to-tape or tape to cable joints. Wiped joints in aluminium sheathed cables should always be protected against moisture by a suitable paint, by an impervious tape or by embedding in bitumen.

Bare copper sheathed cables should not be laid in contact with zinc plated (galvanized) materials such as cable tray in damp conditions. This is because the electropotential series indicates that zinc is anodic to copper and therefore preferential corrosion of the zinc plating may occur. This action will not affect the copper but may cause corrosion of the cable tray. The presence of moisture is essential to produce electrolytic action, so in dry conditions this action will not occur. If moisture is present then electrolytic action will take place but the extent of any corrosion is dependent upon the relative areas of the two metals and the conductivity of the electrolyte (moisture in this instance). In these circumstances, thermoplastic sheathing should be used.

Hostile environments and chemicals also can attack the conductor, its insulation and sheath and any enclosure or equipment.

A few examples of this are:

- petroleum products, creosote; some solvents and hydrocarbons attack rubber and may attack polymeric materials such as thermoplastic
- plasticisers migrate to polystyrene from thermoplastic and also to some types of plaster (see section 7.5 also)
- hostile atmospheres (e.g. those in the vicinity of plastics processing machinery, and vulcanising, which produces a sulphurous atmosphere)
- coastal areas with salt-laden air
- locations where animals are kept (agricultural environments, kennels, etc.), as animal urine can be corrosive.

In such cases the cable or equipment manufacturer's advice should be obtained and care taken in the measures employed.

To provide adequate protection from corrosion, the corrosive substances need to be identified and the manufacturer's or a specialist's advice obtained.

4.8 Impact (AG), vibration (AH) and other mechanical stresses (AJ)

522.6 Any part of the fixed installation which may be exposed to mechanical damage must be able to survive it. Typical examples of mechanical damage will include impacts from would be outside the scope of BS 7671) the following must be followed:

422.1.1 **(a)** restrict electrical equipment to only that necessary for the location

(b) select only equipment appropriate for the location, e.g. of suitable IP rating

422.3.4 **(c)** where wiring is not completely embedded in non-combustible material, such as plaster or concrete, select wiring systems with non-fire propagation characteristics, e.g. cables enclosed in conduit complying with the non-flame propagating requirements of BS EN 61386-21 or ducting or trunking complying with BS EN 50085 or mineral insulated sheathed cable (MICC). Cable types complying with BS EN 60332-1-2 are acceptable

422.3.5 **(d)** if, of necessity, a wiring system traverses such a location but is not necessary to the use of the location, then the cabling should have no connections unless in an enclosure meeting the fire test requirements of BS 6458-2.1 (IEC 60695-2-1:1980) with the following test temperatures:
(i) 750 °C for surface mounted installations
(ii) 850 °C for installation in combustible hollow walls

422.3.10 **(e)** place overload and short-circuit protective devices upstream and outside the location

422.3.9 **(f)** except for mineral insulated cables and busbar trunking or powertrack systems, fit an RCD with a rated residual operating current of 300 mA or less. Where there is concern that quite low current faults could cause fire, the rated residual operating current should not exceed 30 mA

422.3.12 **(g)** provide all wiring systems with a protective conductor. PEN conductors must not be installed for equipment within the location

422.3.13 **(h)** install an isolating device (disconnector) outside the location to switch off all supplies to the location

(i) always switch the neutral as well as the line conductors when isolating equipment

(j) don't install bare live conductors for sliding contact wiring except where there is no practical alternative, e.g. sliding contacts for a hoist when either an enclosed system or flexible cables should be used

422.3.11 **(k)** if SELV or PELV circuits are installed, contain all live parts within enclosures to at least IP2X or provide insulation

422.3.7 **(l)** protect all motors (that are not continuously supervised) against excessive temperature by overload protective devices with manual resetting or equivalent

422.3.7 **(m)** protect motors with:
(i) star-delta motors – protection monitoring in both star and delta configurations
(ii) slipring starters from being left with resistance in the rotor circuit

422.3.2 **(n)** use luminaires with limited surface temperatures only. Fittings manufactured to
422.3.8 BS EN 60598 have surface temperatures limited to 90 °C under normal conditions and 115 °C under fault

422.3.1 **(o)** avoid the use of small spotlights and projectors. Any that are necessary should be separated from combustible materials by minimum distances as follows:
(i) up to 100 W – 0.5 m
(ii) over 100 W up to 300 W – 0.8 m
(iii) over 300 W up to 500 W – 1 m
Note: Greater distances may be necessary if the lamps are focused on combustible material.

422.3.8 **(p)** take precautions against components that are likely to run hot, such as lamps, from falling out of luminaires

422.3.203 **(q)** fix all heating appliances, and for storage types select only ones which will
422.3.202

prevent the ignition of combustible dust or fibres

422.3.2 **(r)** ensure by selection that, in normal operation, enclosures of thermal appliances such as heaters do not attain a temperature higher than 90 °C.

422.4 **4.11.2 Locations constructed of combustible materials**

Electrical equipment which is to be installed on or in a combustible wall must:

422.4.201 **(a)** comply with the relevant standard. There is a requirement on manufacturers to ensure that electrical equipment meets the fire resistance tests within the standards which cover their products; or

422.4.202 **(b)** be enclosed with a suitable thickness of non-flammable material. As such materials generally provide excellent heat insulation, this could have the effect of making the interior of the enclosure hotter where flush-mounted equipment is separated from the building material by additional thermal insulation.

Cables

422.4.203 Cables must comply with the flame tests in BS EN 60332-1-2:2004 *Tests for vertical flame propagation. Procedure for 1 kW pre-mixed flame*, which are carried out by manufacturers.

Cables manufactured to the following standards all meet the flame tests:

BS 5467, BS 6004, BS 6346, BS 6724, BS 6883, BS 7211, BS 7629, BS 7846, BS 7889 and BS 7919.

See Table F1 in Appendix F for a comparison of harmonized cable types to BS 6004.

Conduit and trunking systems

422.4.204 Conduit systems must be in accordance with BS EN 61386-1 *Conduit systems. General requirements*. Trunking systems and ducting systems must be in accordance with BS EN 50085-1 *Cable trunking and cable ducting for electrical systems. General requirements*. There is a requirement on manufacturers to ensure that conduit, trunking and ducting systems meet the fire resistance tests within these standards.

Further information on both standards is contained in section 5.1 of this Guidance Note.

4.11.3 Fire propagating structures

422.5.1 In structures or buildings where the shape and dimensions are such as will facilitate the spread of fire (e.g. chimney effect), precautions are required to be taken to ensure that the electrical installation cannot propagate a fire.

Sect 527 Suitable cables and wiring systems must be installed and a high standard of sealing around and, as appropriate, within wiring systems are essential to minimise the risk of propagation of fire.

4.11.4 Locations of national, commercial, industrial or public significance

422.6 Regulation 422.6 states that the requirements of Regulation 422.1 shall apply to
422.1 locations that include buildings or rooms with assets of significant value and gives examples that include national monuments and museums. Other examples given of buildings having commercial, industrial or public significance are railway stations and

between fire zones or compartments without further physical fire protection, unless there is significant risk of fire e.g. hazardous or flammable materials stored or special processes. Cables not to BS EN 60332-1-2 may not pass unprotected between building fire compartments or zones and should generally only be used in short lengths and not in areas of fire risk.

422.3.4 Most types of general cable comply with BS EN 60332-1-2, including thermoplastic insulated and sheathed cables to BS 6004 and thermosetting/SWA/thermoplastic cables to BS 5467. A more severe test is prescribed in the BS EN 60332-3 series, which tests flame propagation of a group of cables, and it may be that cables installed in groups in risers, or control and communications cables where bunched, should be required to comply with this standard.

521.9.1 Flexible cables may be used for fixed wiring and should generally be of the heavy duty type.

5.1.1   Conduit, trunking and ducting systems standards

The BS EN 61386 series of standards covering conduit systems for electrical installations has replaced the following earlier standards:

▶ BS 731-1:1952 (1993) *Flexible steel conduits*
▶ BS 4568 *Steel conduit and fittings with metric threads*
▶ BS 4607 *Non-metallic conduit and fittings*
▶ BS 6099 *Conduits for electrical installations*
▶ BS EN 50086-1:1994 *Specification for conduit systems for cable management. General requirements*
▶ BS EN 50086-2-1:1996 *Specification for conduit systems for cable management. Particular requirements. Rigid conduit systems*
▶ BS EN 50086-2-2:1996 *Specification for conduit systems for cable management. Particular requirements. Pliable conduit systems*
▶ BS EN 50086-2-3:1996 *Specification for conduit systems for cable management. Particular requirements. Flexible conduit systems.*

The following relevant parts of the BS EN 61386 series have been published:

▶ BS EN 61386-1:2008 *Conduit systems for cable management. General requirements*
▶ BS EN 61386-21:2004 *Conduit systems for cable management. Particular requirements. Rigid conduit systems*
▶ BS EN 61386-22:2004 *Conduit systems for cable management. Particular requirements. Pliable conduit systems*
▶ BS EN 61386-23:2004 *Conduit systems for cable management. Particular requirements. Flexible conduit systems*
▶ BS EN 61386-24:2010 *Conduit systems buried underground.*

The latest standards should always be used whenever possible.

The designer must select an appropriate level of performance for application for the specific characteristics of the installation and, again, overspecification in design will increase costs. A listing of performance criteria for conduit is given in Table 5.1.

▼ **Table 5.1**   Performance requirements in BS EN 61386-1 Conduit systems for cable management. General requirements

Performance characteristics	Performance classification							
	0	1	2	3	4	5	6	7
Compression (N)	0	125	320	750	1250	4000	–	–
Impact energy (joules)	–	4.9	9.8	20	59	196	–	–
Low temperature (°C)	–	+5	−5	−15	−25	−45	–	–
High temperature (°C)	–	+60	+90	+105	+120	+150	+250	+400
IP rating								
Solid object	–	–	–	2.5 mm	1.0 mm	dust protected	dust tight	–
Water	0	drops	drops at 15°	spray	splashing	jets	powerful jets	temporary immersion
Corrosion protection	–	low inside and out-side	medium inside and outside	medium inside, high out-side	high inside and out-side	–	–	–
Tensile strength (N)	0	100	250	500	1000	2500	–	–
Suspended load (N)	0	20	30	150	450	850	–	–

527.1.6 Specific flame propagation tests are included in the conduit standards. Flame propagation is defined in terms of a pass or fail test based on temperatures consistent with human safety (temperatures above which persons cannot survive). Materials that pass the test are identified as 'non-flame propagating' and are required to be marked with a clearly legible identification or may be of any colour except yellow, orange or red, unless clearly marked on the product to be of non-flame propagating material. Flame propagating conduit must be coloured orange.

The BS EN 50085 series of standards cover cable trunking systems and cable ducting systems; the following relevant parts have been published:

▶ BS EN 50085-1:2005 *Cable trunking systems and cable ducting systems for electrical installations. General requirements*
▶ BS EN 50085-2-1:2006+A1:2011 *Cable trunking systems and cable ducting systems for electrical installations. Cable trunking systems and cable ducting systems intended for mounting on walls and ceilings*
▶ BS EN 50085-2-2:2008 *Cable trunking systems and cable ducting systems for electrical installations – Part 2-2: Particular requirements for cable trunking systems and cable ducting systems intended for mounting underfloor, flushfloor or onfloor*
▶ BS EN 50085-2-3:2011 *Cable trunking and cable ducting systems for electrical installations. Particular requirements for slotted cable trunking systems intended for installation in cabinets. Section 3: Slotted in cabinets.*

the requirements of the Regulations for a protective conductor.

(c) Be enclosed in earthed trunking or ducting complying with BS EN 50085-2-1 and satisfying the requirements of the Regulations for a protective conductor.

(d) Be provided with mechanical protection against damage sufficient to prevent penetration of the cable by nails, screws and the like.

Note: Unless purpose manufactured plates are used, effective protection is very difficult to achieve, bearing in mind modern fixings such as self-tapping screws and shot-fired nails that are available.

(e) Form part of a SELV or PELV circuit meeting the requirements of Regulation 414.4.

▼ **Figure 5.2**   Holes and notches in joists

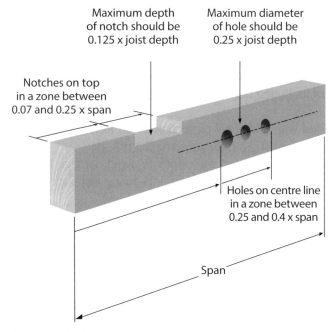

Maximum depth of notch should be 0.125 x joist depth

Maximum diameter of hole should be 0.25 x joist depth

Notches on top in a zone between 0.07 and 0.25 x span

Holes on centre line in a zone between 0.25 and 0.4 x span

Span

Note: Notches and drillings in the same joist should be at least 3 diameters apart horizontally.

▼ **Figure 5.3**   Cables in floor joists

cable in earthed steel conduit

insulated and sheathed cable

greater than 50 mm

joists

5.3   Cable routes and livestock

Sect 705   Particular attention is needed to the routing of cables in locations where livestock is present. (See section 4.9 also.)

Animals in enclosures or buildings may rear up or move as a herd. Such movement can result in severe damage to inappropriately positioned equipment and risks to the animals themselves. Animals may also create corrosive conditions so that normal finishes are not durable enough. Heavy duty equipment with anti-corrosive finishes should be selected and appropriately sited.

Heavy duty high impact PVC equipment with suitable Class II accessories and conduit and enclosures have been found to be satisfactory in many cases.

5.4   Capacities of conduit and trunking

522.8   Where cables are installed in conduit or trunking, care is needed to prevent undue
Appx A   strain when installing the cables. Appendix A of this publication provides information on the size of conduit or trunking necessary to accommodate cables of the same size or mixed sizes.

It should be remembered that cables installed in conduit or trunking must be sized to allow also for the thermal effect of the grouping of such cables. Where advantage is taken of rating factors it may be necessary to indicate this by a local label in the trunking or a note in the installation circuit chart to warn future modifiers. Care must be taken when drawing cables into conduit and trunking systems to avoid damage by abrasion.

5.5   Selection of cables and wiring systems with low emission of smoke and corrosive gases when affected by fire

When specifying or selecting cables in areas with high levels of public access, for example hospitals, schools, supermarkets and shopping developments, the designer should consider factors which may affect the safety of the occupants, such as:

(a) where the rapid formation of dense, noxious smoke would be likely to produce injury or panic in individuals or crowds
(b) along escape routes, where it is important that the rapid but orderly exit of people is not impeded by smoke or fumes
(c) where the formation of corrosive halogen-acid products would have a harmful effect upon expensive equipment, i.e. IT equipment.

Appx 1   A list of British Standards is given in Appendix 1 of BS 7671 and includes reference to cables with low emission of smoke and corrosive gases when affected by fire. These include BS 6724, BS 7211, BS 7629, BS 7846 and BS 8436.

Appx 5   The designer should take into consideration the classification BD1, BD2, BD3 or BD4
Sect 422   according to Appendix 5 of BS 7671 and liaise with the individual or organisation responsible for the fire safety of the building to specify one of these where they have not done so.

Unplasticised thermoplastics, as used in the manufacture of many cable management products (such as conduit, ducting, trunking and cable tray), are non-flame propagating but are not low smoke zero halogen (LSOH) unless they have been specifically ordered as low smoke zero halogen.

It is not expected that general electric cables in the quantities installed in most buildings would provide high levels of smoke on their own and cables would usually only be expected to burn when other building components are also on fire. However, in some installations, such as public buildings and hospitals, it is preferable to take positive action to limit the production of smoke and fumes. Cables with low emission of smoke and corrosive gases when affected by fire have an important place in the installation design.

Tests for smoke and acid gas emission identify the performance of cables in fire conditions. These tests need to take into account the combination effect of all the materials used in a wiring system and environmental conditions. Specific requirements for smoke and acid gas limits must be specified by the designer and statutory authorities where this is relevant. It must be realised that needless overspecification of cables with low emission of smoke and fumes will impose an unnecessary cost burden on a project.

5.6 'Section 20'

Buildings within inner London are required to comply with the special provisions of 'Section 20' and these rules may be extended to apply to other cities at the discretion of the local authority.

Section 20 of the London Building Acts (Amendment) Act 1939 (as amended primarily by the Building (Inner London) Regulations 1985) is principally concerned with the danger arising from fire within certain classes of building which by reason of height, cubic capacity and/or use necessitate special consideration. The types of building coming within these categories are defined under Section 20 of the amended 1939 Act.

As buildings vary so much in height, cubic capacity, layout, siting, use and construction, the relevant Council will deal with each case on its merits. The principles seek, in certain buildings (or parts of buildings), to ensure the safety of the structure against fire.

Section 20 applies where:

(a) a building is to be erected with a storey or part of a storey at a height greater than:
  - ▶ 30 m, or
  - ▶ 25 m if the area of the building exceeds 930 m²

(b) a building of the warehouse class, or a building or part of a building used for the purposes of trade or manufacture, exceeds 7,100 m³ in extent unless it is divided by division walls in such a manner that no division of the building is of a volume exceeding 7,100 m³.

A fire alarm system complying with the current edition of BS 5839-1 should be provided throughout every building except buildings comprising flats and/or maisonettes. These are covered by BS 5839-6 and also may have specific Building Regulations requirements.

In some instances, it may be necessary for such a fire alarm installation to be automatic where the use of the building (or part) warrants it (e.g. in hotels).

In office buildings where the means of escape is based on phased evacuation a number of additional features (such as a public address system) will be necessary.

The fire and rescue service when responding to an emergency in a building should have the indicator panels and associated manual controls for the building's fire protection systems located together in one place designated as the fire control centre.

Smoke extraction should be provided from each storey by openable windows or by a mechanical smoke extraction system.

Diesel engine generators or pumps should be enclosed by walls and a roof, of non-combustible construction having a standard of fire resistance of not less than four hours. Any openings in the walls should be protected by a single self-closing 120-minute door.

Oil storage in connection with the foregoing should comply with the fire authority requirements.

Transformer substations and switchrooms containing electrical oil-cooled transformers or oil-filled switchgear with an oil capacity in excess of 250 litres will be required to be housed in a fire-resistant substation area and cast resin, or dry-type transformers are preferred.

Conditions will normally be imposed requiring details of the general heating, lighting, electrical and ventilation arrangements to be submitted to Building Control for approval before commencement of the works and such works to be provided and maintained to the satisfaction of Building Control. Periodic inspections may be made by local Building Control inspectors of the approved heating, lighting, electrical and ventilation installations during installation.

The foregoing Section 20 requirements apply specifically to London but other large cities have their own similar requirements.

5.7 Buried cables

522.8.10 Cables to be installed in ducts or pipes in the ground are not required by BS 7671 to have armouring and/or a metal sheath, if the duct or pipe is sufficiently strong to resist likely mechanical damage. In case of doubt, a cable fulfilling the requirements of Regulation 522.8.10 should be provided.

542.3.1 Protective conductors which are buried directly in the ground must comply with the requirements of Regulation 542.3.1.

522.8.10 All buried cables are required to be marked by cable covers or a suitable marker tape and buried conduits or ducts must also be suitably identified. A warning tape above the duct or conduit run would be suitable.

Cables, conduits and ducts need to be buried deep enough to avoid being damaged by any disturbance of the ground reasonably likely to occur. This could include gardening/ horticultural works, excavating for any local buried service pipes, etc. Generally, it is better to locate cables, etc., clear of such possible works.

A depth of burial of less than 0.5 m is usually inadvisable as shallow laid cables may be inadvertently damaged by general gardening, etc. Cables that cannot be buried at a reasonable depth should be specifically protected, e.g. by ducts encased in concrete, or installed along an alternative route.

Before any excavation is undertaken for cable or other works HSE guidance booklet HSG47 Avoiding danger from underground services should be studied, as it provides valuable advice on safety aspects.

705.522   In areas of agricultural and horticultural premises where vehicles and mobile agricultural machines are operated, cables must be buried in the ground at a depth of at least 0.6 m, with added mechanical protection. Cables in arable or cultivated ground must be buried at a depth of at least 1 m.

708.521.1.1   In caravan parks, campgrounds and similar locations, cables must be buried at a depth of at least 0.6 m and, unless having additional mechanical protection, be placed well outside any caravan pitch or away from areas where tent pegs or ground anchors are expected to be hammered into the ground.

709.521.1.7   In marinas the recommended minimum buried depth is 0.5 m.

522.8.10   It is important to be able to identify exactly where hidden services are located and accurate records, including drawings, should be made before trenches are backfilled. Cable routes should be located on drawings by dimensions from some fixed object that is not expected to change for a considerable time (e.g. a building). Road and pathway features may be relatively easily changed. Scale drawings are of little value unless they are detailed. Concrete route markers may be installed at changes of direction of cable routes and at regular intervals along routes. It may also be useful to lay cable warning marker tape just below ground level along cable routes so this will be exposed at any future digging well before cables are accessed or disturbed.

The Electricity Safety, Quality and Continuity Regulations 2002 require that all buried cables shall be marked or otherwise protected (Regulation 14(2)).

Many public service undertakings with services buried along the public highway have agreed in the National Joint Utilities Group (NJUG) a specific colour identification system for their service or ductwork. Generally, each of the Statutory Undertakings has been allocated a specific colour for their underground services. Details are given in Tables 5.2 and 5.3.

In addition, Local Authorities and, in particular, Highway Authorities have allocated themselves colours appropriate for their own installations. These vary from area to area and on occasion may clash with the NJUG colour code. In the interests of safety, Local Authorities should always be asked to supply details before excavation is commenced in their area.

The colour code is not universally applied, as many of the older services were installed before this allocation was developed. Other means of identifying cables include clay or concrete tiles laid above electricity cables, usually with the word 'electricity' embossed into the top surface. The use of coloured plastic tape or board which would have the details of the service printed on its surface is a more recent method.

522.8.11   Where cables pass through holes in metalwork, brickwork, timber, etc., precautions must be taken to prevent damage to the cables from any sharp edges, or alternatively mechanical protection should be provided, sufficient to prevent abrasion or penetration of the cable.

▼ **Table 5.2** NJUG recommended depths of underground general utility apparatus

Asset owner	Duct	Pipe	Cable	Marker systems	Recommended minimum depths (mm)	
					Footway	Carriageway
Electricity HV (High Voltage)	Black or red duct or tile	N/A	Black or red	Yellow with black and red legend or concrete tiles	450–1200	750–1200
Electricity LV (Low Voltage)					450	600
Gas	Yellow	*** See row below	N/A	Black legend on PE pipes every linear metre	600 (footway) 750 (verge)	750
	*** PE – up to 2 bar – yellow or yellow with brown stripes (removable skin revealing white or black core pipe).					
	– between 2 to 7 bar – orange.					
	Steel pipes may have yellow wrap or black tar coating or no coating.					
	Ductile Iron may have plastic wrapping.					
	Asbestos & Pit / Spun Cast Iron – No distinguishable colour.					
Water – non-potable & grey	N/A	Black with green stripes	N/A	N/A	600–750	600–750
Water – firefighting	N/A	Black with red stripes or bands	N/A	N/A	600–750	600–750
Oil / fuel pipelines	N/A	Black	N/A	Various surface markers Marker tape or tiles above red concrete	900 *All work within 3 metres of oil fuel pipelines must receive prior approval*	900 *All work within 3 metres of oil fuel pipelines must receive prior approval*
Sewerage	Black	No distinguishing colour/ material (e.g. Ductile Iron may be red; PVC may be brown)	N/A	N/A	Variable	Variable
Telecomms	Grey, white, green, black, purple	N/A	Black or light grey	Various	250–350	450–600
Water	Blue or grey	Blue polymer or blue or uncoated Iron / GRP. Blue polymer with brown stripe *(removable skin revealing white or black pipe)*	N/A	Blue or blue/ black	750	750 min
Water pipes for special purposes (e.g. contaminated ground)	N/A	Blue polymer with brown stripe *(non-removable skin)*	N/A	Blue or blue/ black	750	750 min

Sizing of cables 6

6.1   Current-carrying capacity and voltage drop

Appx 4   The preface to the current rating and voltage drop tables of Appendix 4 of BS 7671 advises how they should be used to determine the cross-sectional area of a conductor. This will depend, among other factors, on the type of overcurrent protection provided. Guidance Note 6 gives further information on cable sizing.

The design current, I_b, of a circuit should be that drawn by the current-using equipment (load) at the nominal voltage of the supply.

An accurate value of design current is required in order that cable sizing can be correctly carried out.

The basic formulae to obtain a design current are as follows:

Single-phase load,

$$I = \frac{kW}{V \times p.f.} \times 1000 \ (A)$$

(6.1)

Three-phase load,

$$I_l = \frac{kW}{\sqrt{3} \, V_l \, p.f.} \times 1000 \ (A)$$

(6.2)

where:

V is the nominal line (phase) voltage to Earth, also denoted U_0
V_l is the line-to-line voltage, also denoted U
p.f. is the power factor
I_l is the line current in a three-phase system.

Table 52.3   Table 52.3 of BS 7671 considers only the minimum cross-sectional area of conductor that should be used. The actual cross-sectional area chosen depends on a number of factors, which are explained in the preface to the tables of Appendix 4 of BS 7671.

552.1.1   Some loads such as motors, transformers and some electronic circuits take large surge (inrush) currents when they are switched on. Due account must be taken of any extra cumulative heating effect due to motor inrush currents or electric braking current where the motors are for intermittent duty with frequent stopping and starting.

6.2   Diversity

311.1   Diversity at any point in time is the actual load (or estimated maximum demand) divided by the potential load (or connected load), usually expressed as a percentage. The load that may be taken for the basis of selection of cable and switchgear capacities,

525.203 **Notes:**

(a) A greater voltage drop may be acceptable for a motor circuit during starting and for other equipment with a high inrush current, provided that in both cases it is ensured that the voltage variations remain within the limits specified in the relevant equipment standard.

(b) The following temporary conditions are excluded:
- voltage transients
- voltage variation due to abnormal operation.

Voltage drops may be determined from the data given in Appendix 4.

The Electricity Safety, Quality and Continuity Regulations 2002 allow a supply variation of +10 % to −6 % .

When designing large and intermediately sized installations, it can help to apportion the installation's overall voltage drop into smaller percentages allocated to parts of the installation. For example, the designer can apportion a design voltage drop of 4% to the final circuits and two per cent voltage drop to the sub-mains distribution. This method has proven to be successful in providing both economical and functional designs.

Other influences 7

7.1 Electrical connections

522.5
526.1
526.2
Connections between cables and other equipment need to be selected with care. There is a risk of corrosion between dissimilar metals, depending upon the environment. The manufacturer's recommendations must be observed and the Standard followed. The power rating and the physical capacity of terminals must be adequate, as must their mechanical strength, when terminating larger conductors.

Thermosetting insulated cables to BS 5467 and thermoplastic cables to BS 6004 can operate at higher conductor temperatures than other thermoplastic insulated cables. The higher conductor temperature results in an increased current-carrying capacity (10–20 per cent) and consequently higher voltage drop values.

512.1.5
However, the use of cables other than standard 70 °C thermoplastic cables requires careful selection of termination systems. Where cables exceeding this temperature rating are selected, for example 90 °C thermosetting cables, the designer or installer must select equipment designed for the possible increased temperature at the terminations. Alternatively, the designer or installer could size the cable as if it were a thermoplastic cable, ensuring that the conductor temperature will not exceed 70 °C in normal conditions.

522.2.201
Higher temperature cable may be a necessity where the heat of the equipment to which it is to be connected may be excessive. Thermoplastic or thermosetting conductors used in this way may need their insulation locally replaced or supplemented by an insulation with a more suitable temperature rating (e.g. at an enclosed luminaire). Where it is necessary to adjust the cross-sectional area of a cable at the point of termination this should be done by the use of a recognised technique.

421.1.6
526.3
Any enclosure must have suitable mechanical and fire resistance properties. Unless exempt, joints must remain accessible for inspection and maintenance.

526.2
526.6
Account must be taken of any mechanical damage and vibration likely to occur. The method of connection adopted should not impose any significant mechanical strain on the connections, nor should there be any mechanical damage to the cable conductors.

Cable glands, clamps and compression-type joints should retain securely all the strands of the conductor. Every compression joint should be of a type which has been the subject of a test certificate as described in BS EN 61238-1:2003, and the appropriate tools and methods specified by the manufacturer of the joint connectors should be used. With the exception of protective conductors greater than 2.5 mm^2, a termination or joint in an insulated conductor should be made in an accessory or luminaire complying with the appropriate British Standard. Where this is not practicable, it should be enclosed in material designated non-combustible when tested to BS 476-4 or complying with 'P' to BS 476-12 or the relevant requirements of BS 6458, Section 2.1.

Such an enclosure may be formed by part of an accessory and/or luminaire or by a part of the building structure.

526.5
526.8
All connections, including those of ELV 12 V luminaires, must be enclosed in accordance with the requirements of Regulation 526.5. The use of unenclosed terminal blocks or terminal strips wrapped within insulating tape does not comply with the requirements of the Regulations. Connections must be enclosed within a properly constructed junction box or enclosure, either to the relevant British Standard (e.g. BS 4662 or BS EN 60670) or made to an equivalent flame propagation which passes the glow-wire test of BS EN 60695-2-11. All insulated, unsheathed cable, such as cable ends stripped back, must be contained within the enclosure.

Terminations of mineral insulated cables should be provided with sleeves having a temperature rating similar to that of the seals. Cable glands should securely retain without damage the outer sheath or armour of the cables. Mechanical cable glands for thermosetting and plastics insulated cables should comply with BS 6121 where appropriate.

553.2
704.511.1
Appropriate cable couplers should be used for connecting together lengths of flexible cable. Generally cable couplers should be non-reversible and to an appropriate Standard. On construction sites, where the requirements apply, every plug and socket-outlet of rating 16 A or greater must comply with BS EN 60309-2 (formerly BS 4343).

Conduit should be free from burrs and swarf and all ends should be reamed to obviate damage to cables. Substantial boxes of ample capacity should be provided at every junction involving a cable connection in a conduit system. The designer/installer should be aware that large cables need a greater factor of space for installation and removal.

537.2.1.7
Regulation 537.2.1.7 requires a provision for disconnecting the neutral conductor for isolation or testing purposes. A suitable joint will be adequate for smaller cables but a bolted link could be used for larger conductors due to the physical effort necessary to move the cables for disconnection/reconnection.

526.3
Every electrical connection must be accessible for inspection with the exception of connections that are considered robust or maintenance-free as follows:

▶ joints designed to be buried in the ground
▶ an encapsulated joint
▶ a compound filled joint
▶ a cold tail of a heating connection
▶ welded or soldered joints
▶ joints made using appropriate compression tools
▶ certain spring-loaded terminals complying with BS 5733 and marked with the symbol ⓂⒻ.

7.2   Cables in contact with thermal insulation

523.9
Where a cable is to be run in a space in which thermal insulation is likely to be placed, the cable should, wherever practicable, be fixed in a position such that it will not be in contact with or covered by the thermal insulation. Surrounding the cable totally with thermal insulation will mean derating the cable by a factor of 0.5 (except for short lengths of cable, see below).

Table 4D5   For lofts and intermediate floor structures with thermal insulation, a dedicated current-carrying capacity table is given in BS 7671, Table 4D5. By installing cables to 'installation method' 100 or 101 of Table 4A2 then the 'standard' domestic cable sizes can continue to be used in these loft and floor situations.

Figure 7.1 indicates installation method 100 of Table 4A2.

▼ **Figure 7.1**   Cables in thermal insulation, installed as per installation method 100 of Table 4A2

Appx 4   For a cable installed in a thermally insulated wall, the cable being in contact with a thermally conductive surface on one side, current-carrying capacities are tabulated in Appendix 4 of BS 7671, Method A being the appropriate reference method.

For a single cable likely to be totally surrounded by thermally insulating material over a length of more than 0.5 m, the current-carrying capacity should be taken, in the absence of more precise information, as 0.5 times the current-carrying capacity for that cable clipped direct to a surface and open (Reference Method C).

Table 52.2   Where a cable is totally surrounded by thermal insulation for less than 0.5 m the current-carrying capacity of the cable must be reduced appropriately depending on the size of cable, length in insulation and thermal properties of the insulation. The derating factors in Table 7.1 are appropriate to conductor sizes up to 10 mm^2 in thermal insulation having a thermal conductivity greater than 0.04 Wm^{-1}K^{-1}.

▼ **Table 7.1** Cables surrounded by thermal insulation (Table 52.2 of BS 7671)

Length in insulation (mm)	Derating factor
50	0.88
100	0.78
200	0.63
400	0.51

The above table is based upon normal cavity insulation such as 'Rockwool'. More precise data can be obtained from ERA Report 85-0111 *The temperature rise of cables passing through short lengths of thermal insulation.*

Typical cable sizes

Appx 4   For domestic lighting circuits, the load does not normally exceed 6 A. The smallest
433.1.1(ii)   cable used will be 1 mm^2 with a current-carrying capacity for Method C of 16 A (Table 4D5, column 6). If totally surrounded by thermal insulation for 0.5 m or a greater

length the current-carrying capacity can therefore be taken as 8 A, which would be acceptable with a 6 A overcurrent protective device.

433.1.1(ii) A cable sometimes used for cooker circuits is 6 mm^2, which has a current-carrying capacity of 47 A when clipped direct, reducing to 23.5 A if it is installed in thermal insulation. A 30 A or 32 A overcurrent device is usually employed. The current taken by a cooker varies over a wide range and theoretical loads in excess of 23.5 A can be expected. This would lead to overheating of the cable if continuously carrying such current. Hence, a larger cable may be necessary. Also, the rated current (I_n) of a protective device must not exceed the lowest of the current-carrying capacities of any of the circuit conductors (I_z).

The loading of cables feeding socket-outlets is even more varied and depends on the type of circuit used, i.e. ring or radial, the number of socket-outlets fed from the circuit and the type of load. For example, 2.5 mm^2 thermoplastic insulated and sheathed cable used in a multi-socket-outlet radial circuit protected by a 20 A overcurrent protective device has a 'clipped direct' current-carrying capacity of 27 A (Table 4D2A, column 6), reducing to 13.5 A if the cable is embedded in thermal insulation.

When clipped direct the cable could supply up to 6.2 kW without overheating, though because, in this example, the circuit is protected by a 20 A device it should not be loaded to more than 4.6 kW in any event.

Loading as great as this in domestic installations is infrequently encountered and much more common are small loads such as standard lamps and television sets, up to about 1 kW, i.e. 4 A in total.

However, if the cable is later embedded in thermal insulation, the reduced current-carrying capacity of 13.5 A would be exceeded if a load of 3.2 kW or greater were connected and overheating would be a possibility.

Thus, the introduction of thermal insulation into cavities where cables are already installed will create the possibility of overheating the cable. The extent of the risk will depend on the type of thermal insulation used and the total loading of the cables.

7.3   Mutual or individual deterioration

522.5.3 BS 7671 requires that materials liable to cause mutual or individual deterioration or hazardous degradation shall not be placed in contact with each other. One popular type of thermal insulation for cavity walls is urea formaldehyde foam and there should be no adverse reaction between this material and thermoplastic.

Other materials may be used for cavity insulation but the supplier does not always disclose their chemical composition. Before allowing any thermal insulation materials to come into contact with cable insulation or sheath materials the thermal insulation supplier should be asked to confirm in writing that there will be no adverse effect on the cable insulation or sheath. Where there is doubt, an inert barrier should be inserted between the cable and the thermal insulation.

Expanded polystyrene sheets or granules are used in construction (e.g. insulating lofts and supporting floors) and may be used in cavity walls. If this material comes into contact with thermoplastic cable sheathing, some plasticiser will migrate from the PVC to the polystyrene. The thermoplastic will become less flexible and sticky on the surface and the polystyrene will become soft and shrink away from the thermoplastic sheathed cable if possible. Such contact between thermoplastic and polystyrene should be avoided. These comments do not apply to unplasticised PVC conduit and trunking systems.

As only the plasticiser is removed from the sheath, it is not expected that the insulation resistance of the cable will be affected if the cable is not disturbed but this cannot be guaranteed. The cable should be supported away from the polystyrene and a test should be carried out. The cable condition should also be monitored over time in regular periodic inspections. There is no cure for this migration other than replacement of the cable.

Similar remarks apply to some fittings which are in contact with cables. In addition to polystyrene and expanded polystyrene, acrylonitrile-butadiene-styrene (ABS) and polycarbonate are also affected. Nylon, polyester, polyethylene, polypropylene, rigid PVC and most thermosetting plastics are little affected. Natural rubber grommets can become softened but synthetic rubber and PVC grommets are satisfactory.

Contrary to popular belief, plastic insulated and sheathed cables are not suitable for continuous immersion in water, such as in pools and fountains, as they will absorb an amount of water over time and insulation failure will occur. Manufacturers' advice should be taken for applications involving continued immersion of cables.

7.4   Proximity of wiring systems to other services

Sect 528   Section 528 of BS 7671 contains a reference to two voltage bands, equating approximately to ELV and LV. Care must be taken to make sure that only compatible circuits are enclosed in the same conduit or trunking, unless the circuit conductors are insulated for the highest voltage present, or other precautions are taken, such as segregation by a separate compartment or by an earthed metallic screen.

560.7.7   Safety circuits must be separated from other circuits unless they have metallic screens.

For emergency lighting and fire alarms, covered by BS 5266 and BS 5839 respectively, the requirements of the British Standards must be met with respect to segregation.

Where malfunction may occur (e.g. crosstalk with a communication system) all conductors, including the protective conductor, must be correctly routed and adequate separation provided from other cables. A particular form of harmful effect may occur when an electrical installation shares the space occupied by an audio frequency induction loop system – this loop system enables a hearing aid with a telecoil to pick up audio signals from the loop. Under these circumstances, if line and neutral conductors or switch feeds and switch wires are not run close together, 50 Hz interference (and its harmonics) may be picked up by the telecoil of the hearing aid.

Conventional two-way switching systems can often produce inadvertent 50 Hz radiating loops but this can be reduced by arranging the circuits as shown in Figure 7.2.

521.5.1   All the conductors of a circuit should generally follow the same route. Live cables of the same circuit may cause overheating if they enter a ferromagnetic enclosure through different openings.

▼ **Figure 7.2**   Circuit for reducing interference with induction loop

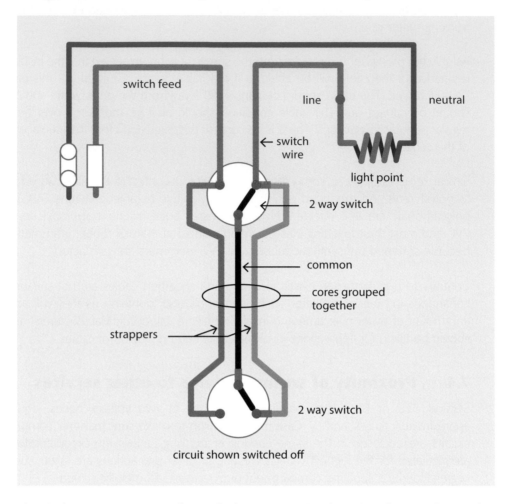

Electrical power systems will usually have immunity from interference from other systems but may cause emission interference with sensitive electronic systems. In addition, harmonic generation and EMC interference will need to be considered, see section 7.6.

7.5   Plasticiser migration from PVC insulation

Installers may occasionally encounter a sticky blue/green deposit in equipment or switch and socket-outlet boxes in older installations wired in thermoplastic insulated cables.

Early manufacture of PVC insulation materials utilized a different formulation from that which is now employed. Thermal cycling of the conductor due to changes in load and conductor temperature drew the thermodynamically unstable insulation plasticiser to the conductor, where, on vertical runs of cable over time, the plasticiser liquid worked down the conductor surface to become a sticky blue/green liquid in the accessory box. The material is not corrosive and can be cleaned away with methylated spirit. The blue/green colouring comes from traces of copper absorbed by the liquid. There is no cure for this plasticiser migration, other than replacement of the cable.

As only the plasticiser is removed from the insulation, it is not expected that the insulation resistance of the cable will be affected if the cable is not disturbed but this cannot be guaranteed and a test should be carried out. The cable condition should also be monitored over time in regular periodic inspections.

Reformulation of PVC compounds in the mid to late 1970s has removed this problem from modern PVC insulated cables.

Migration also happens when bitumen is in contact with PVC. The bitumen absorbs some plasticiser from the PVC. While the amount of loss is insufficient to have much effect on the properties of the PVC, it is enough to cause considerable reduction in the viscosity of the bitumen, which may become so fluid that it can run.

Sect 444

7.6 Section 444 Measures against electromagnetic disturbances

7.6.1 Introduction to electromagnetic compatibility (EMC)

The subject of EMC can be thought of as a series of specifications for equipment and installation rules so that when installed and operated the equipment and installation:

▶ do not emit such high levels of electromagnetic (EM) disturbances as to cause unacceptable electromagnetic interference (EMI) to other equipment within or outside the installation
▶ are sufficiently immune to the EM disturbances in the operating environment that they function well enough.

Note that *equipment* in the above context refers to all electrical, electromechanical and electronic equipment, as well as their system and their installations.

7.6.2 EMC legislation

EMC Regulations 2006

The legal requirement for EMC compliance stems from a European Directive, which is implemented in the UK via the Electromagnetic Compatibility Regulations 2006. This document can be found using the government website www.legislation.gov.uk/uksi and entering the Statutory Instrument (SI) number 3418 for 2006.

Some key points relating to electrical installations from this legislation are as follows:

Responsible person
The EMC Regulations 2006 make a requirement for a responsible person to be appointed for the EMC aspects of an installation. For a new build contract, this responsibility may be that of the main contractor until handover and then it will certainly be the responsibility of the building owner.

332.1

Regulation 332.1 of BS 7671 cuts across any contractual relationship by stating that the installation shall be in accordance with the relevant EMC regulations and with the relevant EMC standard. Electrical designers and installers are recommended to discuss and record the EMC specification and intended equipment to be used by the client.

Essential requirements
The EMC Regulations 2006 make essential requirements which include:

▶ good EMC engineering practices shall be used
▶ equipment shall be installed with regard to its EMC design
▶ the electrical installation method for achieving EMC compliance shall be documented, and
▶ the owner shall retain this EMC documentation for the life of the installation.

7.6.3 Scope of BS 7671 Section 444

Firstly, BS 7671 Part 3 (Assessment of general characteristics) contains the following regulations:

332.1   **332.1** *All electrical installations and equipment shall be in accordance with the EMC regulations and with the relevant EMC standard.*

332.2   **332.2** *Consideration shall be given by the designer of the electrical installation to measures reducing the effect of induced voltage disturbances and electromagnetic interferences (EMI). Measures are given in Chapter 44.*

444.1   Secondly, in Regulation 444.1 the designer is presented with a further requirement:

The designer of the electrical installation shall consider the measures described in Section 444 for reducing electromagnetic disturbances on electrical equipment.

This is similar to Regulation 332.2 and puts the responsibility for considering any additional EMC mitigation measures firmly with the electrical designer.

Section 444 goes on to provide general EMC advice as well as providing EMC bonding solutions and other EMC mitigation techniques; these are discussed later.

In general, the subject of EMC includes harmonic disturbances and overvoltage surges, which are excluded from Section 444 as they are specifically covered in other parts of BS 7671.

7.6.4 Section 444 and general EMC advice

The subject of EMC with application to installations is not straightforward; the key requirement from Section 444 is that the designer should consider the EMC mitigation techniques suggested in the section.

It should be noted that combining two or more CE-marked pieces of equipment does not automatically produce a 'compliant' system, e.g. a combination of CE-marked programmable logic controllers and motor drives may fail to meet the essential requirements of the EMC Regulations 2006. This illustrates the point that the subject of EMC can be a rather complex one.

There are no simple approaches to designing installations for EMC but an example approach is shown in Figure 7.3. The remainder of this section discusses the stages suggested in Figure 7.3 and the advice should be read in parallel with Section 444 of BS 7671.

It is noted that specialist EMC consultants will use many different approaches to EMC design and solutions, including carrying out electromagnetic field strength measurements within completed installations; this latter approach is not always useful for designs that need to be formalised at the early stages of a tender process.

▼ **Figure 7.3** Example of an EMC strategy design approach

a   Sources of electromagnetic disturbances

The electrical power distribution layout and its proximity to the position of any sensitive equipment should be coordinated. For example, it would not be good practice to install a data centre next to an HV intake room, or next to a lift motor room.

444.4.1   Potential sources of electromagnetic disturbances include:

- ▶ switching devices for inductive loads
- ▶ fluorescent lighting and LED lighting
- ▶ electric arc welding units
- ▶ rectifiers
- ▶ choppers
- ▶ frequency convertors/regulators including Variable Speed Drives (VSDs)
- ▶ lifts and escalators
- ▶ power transformers
- ▶ switchgear
- ▶ power distribution busbars
- ▶ powerline communication technology (see below)
- ▶ industrial, scientific and medical (ISM) radio bands (see below).

Powerline communication technology

This technology uses power cables and superimposes a signal or telecommunications data onto the power cable at a 'higher' frequency. The frequency of the superimposed signal can vary widely. For example, lighting control signals may use relatively low frequencies of up to 250 kHz whilst broadband and local area network powerline communication signals use higher frequencies of the order of 28 MHz to 50 MHz as found in the HomePlug type devices. It is noted that HomePlug devices will not provide a good EMC environment, whereas low-data-rate PLC products that comply with BS EN 50065 have been used without any EMI problems being reported.

Industrial, scientific and medical (ISM) radio bands

ISM is a category of the radio spectrum initially intended for industrial, scientific and medical purposes to include items such as microwave ovens and medical diathermy machines. Short-range low power applications also use these frequencies, for

Notes:

1 Screening performance (DC–100 MHz) equivalent to welded mesh steel basket of mesh size 50 mm × 100 mm (excluding ladders). This screening performance is also achieved with steel tray (duct without cover) of less than 1.0 mm wall thickness and more than 20 % equally distributed perforated area.

   No part of the cable within the containment should be less than 10 mm below the top of the barrier.

2 Screening performance (DC–100 MHz) equivalent to steel tray (duct without cover) of 1.0 mm wall thickness and no more than 20 % equally distributed perforated area. This screening performance is also achieved with screened power cables that do not meet the performance defined in Note 1.

   No part of the cable within the containment shall be less than 10 mm below the top of the barrier.

3 Screening performance (DC–100 MHz) equivalent to a fully enclosed steel containment system having a minimum wall thickness of 1.5 mm. Separation specified is in addition to that provided by any divider/barrier.

▶ Achieving zero segregation in the table requires the use of additional segregation/separation for EMC over and above the requirements for safety. Safety considerations must always take precedence over EMC requirements.

As well as this information, Figure 7.5 provides best practices for routing and mounting various types of cable systems in cable management products; see BS EN 50174-2 for more information on this subject, specific to telecommunication installations.

▼ **Figure 7.5**   Best practice for routing configurations of various cable systems

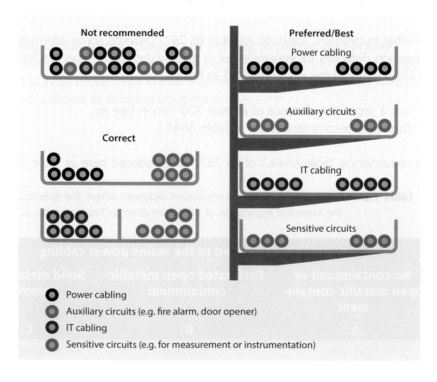

d Select an EMC bonding network appropriate for sensitive equipment

444.4.2.1
A444.1
A444.2
Designers are required to consider the sensitivity of the equipment that will be used in the installation, the location and density of the equipment, the cable routing and EMC sources, and then select an EMC bonding strategy.

Part 2
Table 7.3 and Figure 7.6 provide details of EMC bonding configurations. It is important to note that, with the exception of the star network, the additional bonding is provided purely for EMC mitigation purposes. This additional EMC bonding is defined in BS 7671 as a bonding network (BN) and should be regarded separately to that of the electrical installation safety earthing network or indeed to any safety equipotential bonding requirements. The two systems may share conductors and any conductors or meshes provided for the purposes of EMC mitigation must be connected, at some point, to the installation safety earthing system.

A444.1
Regulation A444.1 suggests that the star network, Figure 7.6A, as traditionally used in many installations, may only be sufficient in domestic and small commercial installations where equipment is not interconnected by signal cables.

For installations with interconnected communication equipment then a multiple meshed solution can be used for smaller installations (see Figure 7.6C) or a full, common meshed bonding network can be provided.

For large commercial installations with concentrated communication data centres and similar, then a common meshed bonding network should be considered, see Figure 7.6D.

A444
The types of EMC bonding networks are detailed in BS 7671 and their relative EMC performance is indicated in Table 7.3. For more information on their application, see BS EN 50310 *Application of equipotential bonding and earthing in buildings with information technology.*

▼ **Table 7.3** EMC bonding configurations and relative performance

Designation	Bonding configuration	BS 7671 definition	Relative EMC mitigation performance
A	star	generally undefined as often EMC network and safety earthing are the same conductors	poorest
B	ring	Bonding ring conductor	better than star
C	multiple meshed	not directly defined but is a small version of the meshed bonding network (see D)	better than star, good for a local solution
D	common meshed	Meshed bonding network	best

The requirements of Section 557 apply to all auxiliary circuits except those covered by a standard for a specific product or system, such as an auxiliary circuit forming part of a switchgear or controlgear assembly manufactured to one or more standards in the BS EN 61439 series.

Part 2 Part 2 of BS 7671 defines an auxiliary circuit as: 'Circuit for transmission of signals intended for control, detection, supervision or measurement of the functional status of a main circuit.'

Examples of auxiliary circuits therefore include circuits for:

▶ remote switching of a circuit via a contactor, such as for emergency switching or for control of lighting,
▶ remote detection of abnormal conditions in a circuit (such a fault current or overload current) via current transformers
▶ remote indication or monitoring of:

  • conditions in a circuit via voltage transformers or current transformers,
  • the status (open, closed or tripped) of a switch or circuit breaker,
  • mains available status,
  • connecting an undervoltage relay to a remote location for monitoring or automatic starting of a generating set,
  • intertripping of circuit-breakers.

7.7.2   Content of Section 557

An appreciation of the content of Section 557 can be gained from Table 7.4, below.

▼ **Table 7.4**   Summary the content of Section 557.

Regulation No	Subject	Remarks
557.1	Scope	Section 557 does not apply to auxiliary circuits covered by specific product or system standards.
557.2	Not used	
557.3	Requirements for auxiliary circuits	
557.3.1	General	Power supply for an auxiliary circuit may be a.c. or d.c. and either dependent or independent of the main circuit. Any status signalling of main circuit must be able to operate independently of that circuit.
557.3.201	Control circuits	Control circuit must be designed, arranged and protected to limit dangers resulting from a fault between control circuit and other conductive parts liable to cause malfunction of controlled equipment.
557.3.2	Power supply for auxiliary circuits dependent on the main circuit	Requirements are given relating to auxiliary circuits with a power supply dependent on the main a.c. circuit, whether directly or via a rectifier or transformer.

Regulation No	Subject	Remarks
557.3.3	Auxiliary circuit supplied by an independent source	Loss of supply or undervoltage of the main circuit source must be detected. An independent auxiliary circuit must not create a hazardous situation.
557.3.4	Auxiliary circuits with or without connection to earth	Certain requirements are given relating to earthing that modify those in other parts of BS 7671.
557.3.5	Power supplies for auxiliary circuits	Requirements are given relating to compatibility of the auxiliary circuit with the supply to the main circuit, nominal voltage of supply to auxiliary circuit, and supplies from batteries.
557.3.6	Protective measures	Requirements for protection against overcurrent, protection against electric shock, and isolation and switching are given, generally reflecting those given elsewhere in Parts 4 and 5 of BS 7671.
557.4	Characteristics of cables and conductors	
557.4.1	Minimum cross-sectional areas	A table of minimum cross-sectional areas of copper conductors is given.
557.5	Requirements for auxiliary circuits used for measurement	
557.5.1	General	
557.5.2	Auxiliary circuits for direct measurement of electrical quantities	Where operation of a fault current protective device could cause danger or lead to a hazardous situation, such operation must also cause disconnection of the main circuit. A device for protection against fault current need not be provided conditions (v) and (vi) of Regulation 434.3 are simultaneous met.
557.5.3	Auxiliary circuits for measurement of electrical quantities via a transformer	A number of requirements relating to where a current transformer is used are given. The secondary side of a voltage transformer must be protected by a short-circuit protective device.
557.6	Functional considerations	
557.6.1	Voltage supply	Where loss or disturbance of voltage could cause auxiliary circuit to be unable to perform its intended function, means must be provided to ensure continued operation of the auxiliary circuit.

▼ **Figure 7.8**   Example of connections to a common conductor

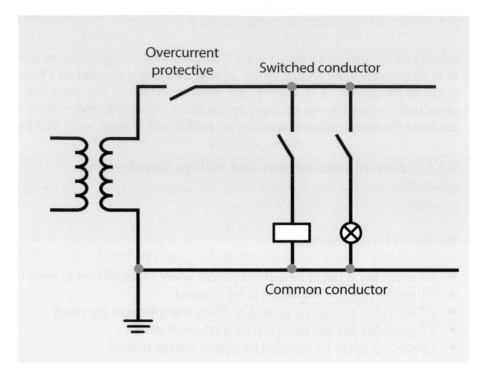

557.6.5.2   For an auxiliary circuits with direct connection to the main circuit between two line conductors, two-pole switching contacts must be used.

557.6.6   Where plug and socket connectors are used for auxiliary circuits, the designer should take measures to prevent unintended disconnection by, for example, the application of a suitable warning label.

Installation of equipment

8

8.1 Earthing requirements for the installation of equipment having high protective conductor currents

8.1.1 Scope

543.7 The requirements of this regulation group apply to:

(a) the connection of current-using equipment having a protective conductor current exceeding 3.5 mA

(b) final circuits and distribution circuits where the accumulated protective conductor current is expected to exceed 10 mA.

The use of information technology (IT) equipment, which has low electrical loading but produces relatively high protective conductor currents, often requires special precautions to be taken. A final circuit can accept many items of IT equipment and remain within the rating of the circuit but the resulting protective conductor currents may be relatively very high.

Modern equipment, such as electronic ballasts in high-frequency fluorescent luminaires or variable-speed drives, is known to create high currents in the protective conductor, hence the precautions of Regulation 543.7 of BS 7671 must be followed.

The levels of expected protective conductor current will vary depending on the type of electrical equipment. BS EN 60335-1 *Household and similar electrical appliances – Safety – Part 1: General requirements* states the fundamental conditions to which items of electrical equipment are manufactured and, hence, should perform when used under normal operating conditions. See Table 11 for permitted maximum protective conductor current values.

Regulation 543.7 contains additional earthing requirements for the installation of equipment having high protective conductor currents. These particular requirements complement the protection described in Section 531 and in Chapters 41 and 54 of BS 7671. Information technology and other electronic equipment usually incorporates filters on the mains conductors. In Class I equipment these filters include capacitors connected between the supply conductors and frame and this results in a standing protective conductor current.

For information technology equipment intended for connection to the supply via a 13 A BS 1363 plug, the maximum allowable standing protective conductor current is 3.5 mA as specified in BS EN 60950. However, modern equipment usually has leakages significantly less than this, perhaps 1 mA to 1.5 mA per item.

With desktop equipment, the use of a number of items of mobile equipment connected to a single socket-outlet or a ring or radial final circuit can result in a cumulative protective conductor current of several milliamperes. This current may be high enough to present danger if the protective conductors become open circuit and, also, may operate an RCD. It should be remembered that an RCD to BS 4293, BS EN 61008 or BS EN 61009 is allowed to operate at any value of residual current between 50 and 100 per cent of its rated residual operating current. Therefore, unwanted tripping could occur under normal running conditions. This is even more likely if there is a surge of protective conductor current when switch-on occurs near the peak value of the supply voltage.

314.1(iv)
531.2.4
Regulation 314.1 therefore requires that the circuit arrangement must be such that the residual current which may be expected to occur, including switch-on surges, will be unlikely to trip the device.

Information technology equipment connected via BS EN 60309-2 (BS 4343) industrial plugs and socket-outlets is permitted to have higher protective conductor currents. However, there is a limit of 10 mA above which the special installation requirements detailed in Regulation 543.7 of BS 7671 apply.

These special requirements are aimed at providing a high-integrity protective conductor. Should the protective conductor of the circuit feeding the equipment fail, the protective conductor current could then pass to earth through a person touching the equipment. As with all installations, the high-integrity protective conductor installation must be periodically inspected and tested to ensure its integrity is maintained (see Guidance Note 3 for further details).

8.1.2   The risks

The risk associated with final circuits having high protective conductor currents is that resulting from discontinuity of the protective conductor, and serious shocks can be received from accessible conductive parts which are not connected to the main earthing terminal of the installation.

This shock risk is extended to all the items of equipment on that particular circuit, whether they individually have high protective conductor currents or not. The more equipment that is connected to a circuit, the wider the risk is spread and the greater the hazard.

IEC publication DD IEC/TS 60479-1 *Effects of current on human beings and livestock* advises on the physiological effects of current passing through the human body. Protective conductor currents exceeding 10 mA can have harmful effects and precautions need to be taken.

543.7.1.202
543.7.1.203
543.7.1.204
The requirements of Regulation 543.7 are intended to increase the reliability of the connection of protective conductors to equipment and to earth, when the circuit protective conductor current exceeds 10 mA. This normally requires duplication of the protective conductor or an increase in its size in the final circuit. The increased size is not to allow for thermal effects of the protective conductor currents, which are insignificant, but to provide for greater mechanical strength and, it is intended, a more reliable connection with the means of earthing. Duplication of a protective conductor, each terminated independently, is likely to be more effective than an increase in size.

8.1.3　Equipment

Note that BS EN 60950, the standard for the safety of information technology equipment, including electrical business equipment, requires equipment with a protective conductor current exceeding 3.5 mA to have internal protective conductor cross-sectional areas not less than 1.0 mm² and also requires a label bearing the following or similar wording fixed adjacent to the equipment primary power connection:

> **HIGH LEAKAGE CURRENT**
>
> **Earth connection essential**
>
> **before connecting the supply**

or

> **WARNING HIGH TOUCH CURRENTS**
>
> **Earth connection essential**
>
> **before connecting the supply**

543.7.1.201
543.7.1.202
543.7.1.203
Regulation 543.7.1.201 requires a single item of equipment, if the protective conductor current exceeds 3.5 mA but does not exceed 10 mA, to be permanently connected or connected by a plug and socket complying with BS EN 60309-2. If the protective conductor current of a single item of equipment exceeds 10 mA the requirements of Regulations 543.7.1.202 and 543.7.1.203 for a high-integrity protective earth connection must be met.

8.1.4　Labelling

543.7.1.205 Distribution boards supplying circuits with high protective conductor currents must be labelled accordingly so that persons working on the boards can ensure they maintain the protective precautions taken (see example of labelling in Figure 8.4).

8.1.5　Ring circuits

543.7.1.204 Ring circuits, by their design, provide duplication of all conductors, including the protective conductor unless this is formed by metallic conduit and/or trunking. Regulation 543.7.1.204 requires that each protective conductor is terminated separately at each connection point; the regulation also requires that each wiring accessory has two separate earth terminals. At socket-outlets, this is a very easy requirement to meet – simply select accessories with dual earth terminals (see Figures 8.1 and 8.2).

▼ **Figure 8.1**　Ring final circuit supplying socket-outlets where the protective conductor current is likely to exceed 10 mA

MET

Ring final circuit supplying equipment with high protective conductor current with ends of each section of protective conductor separately terminated

Earthing conductor

Means of earthing

Accessories must have two terminals for protective conductors. One terminal to be used for each protective conductor.

▼ **Figure 8.2**   Accessory with two terminals for protective conductors

The requirement to terminate duplicate protective conductors separately can pose a problem at the earth bar of the distribution board, i.e. there may not be sufficient terminals to do so. Some large distribution boards may be provided with many spare terminals on the earth bar, while some manufacturers may supply an accessory kit to add extra ways. Where this is not an option, the following example is one method of meeting the requirements.

An electrical installation has a 6-way distribution board/consumer unit and consists of six circuits:

DB 1	
Circuit	**Description**
1	Ring final circuit with high protective conductor current
2	Radial circuit with high protective conductor current
3	Ring final circuit
4	Radial circuit
5	Radial circuit
6	Radial circuit

543.7.1.204   Regulation 543.7.1.204 requires that where two protective conductors are used on circuits with high protective conductor currents, they are to be terminated into separate ways on the earth bar.

By installing circuits 1 and 2 across two ways of the earth bar (see Figure 8.3), in addition to marking each conductor with the correct circuit identification ferrule, there is no ambiguity as to the function of each conductor.

▼ **Figure 8.3**   Termination of circuits with high protective conductor currents at the earth bar

543.7.1.205   Regulation 543.7.1.205 requires that information is provided at the distribution board indicating those circuits having a high protective conductor current. This information could be displayed using the format shown in Figure 8.4.

▼ **Figure 8.4**   Information in accordance with Regulation 543.7.1.205 indicating those circuits with high protective conductor current

> **DB1**
> **The following have high protective conductor current:**
> **1**   Ring final circuit (sales office)
> **2**   Radial final circuit (packaging/dispatch)

8.1.6   Radial circuits

Radial circuits supplying socket-outlets where the protective conductor current is expected to exceed 10 mA are required to have a high-integrity protective conductor connection. This can often be most effectively provided by a separate duplicate protective conductor connecting the last socket directly back to the distribution board, as shown in Figure 8.5. This will provide a duplicate connection for each socket-outlet on the circuit.

The following requirements still apply:

(a) All socket-outlets must have two protective conductor terminals (see Figures 8.2 and 8.5), one for each protective conductor
(b) The duplicate protective conductors must be separately connected at the distribution board (see Figure 8.3).

Note: To reduce interference effects, the duplicate protective conductor should be run in close proximity to the other conductors of the circuit.

▼ **Figure 8.5** Radial circuit supplying socket-outlets (total protective conductor current exceeding 10 mA), with duplicate protective conductor

8.1.7 Busbar systems

543.7.1.203 Busbar systems are often adopted by designers for supplying IT equipment. These may be radial busbars with tee-offs to individual sockets or floor-boxes. The main PE busbar will need to meet one or more of the requirements of Regulation 543.7.1.203:

▶ A cross-sectional area not less than 10 mm^2
▶ A cross-sectional area not less than 4 mm^2 with mechanical protection
▶ Duplicate protective conductors, each of cross-sectional area sufficient to meet the requirements of Section 543.

Tee-offs from a busbar system

If the protective conductor current is expected to be less than 3.5 mA, there is no need for duplication of the protective conductor from the radial busbar to the socket-outlet.

Where the protective conductor current is likely to exceed 10 mA then a single copper protective conductor having a cross-sectional area not less than 4 mm^2 and enclosed to provide additional mechanical protection can be used.

It is necessary for the designer to confirm that the disconnection time of the protective device feeding the busbar will provide adiabatic fault protection to the final conductors.

Overload protection would not be required where the final connection is to a single socket-outlet rated at 13 A, regardless of the size of the protective device for the circuit. However, many floor-box style installations may have up to six socket-outlets per floor-box or branch of the radial circuit and the designer must take this into account when designing such an installation, particularly where the busbar system is protected by a device greater than 32 A rating.

More information on overload protection can be found in Chapter 3 of this publication and in Guidance Note 6: *Protection Against Overcurrent*.

8.1.8 RCDs and high protective conductor currents

411.3.3
415.1.1
314.1 (iv)
531.2.4

To meet the requirements for additional protection, BS 7671 requires that socket-outlets rated at not exceeding 20 A are to be protected by an RCD with an I$_{\Delta n}$ not exceeding 30 mA. Where many items of equipment creating high protective conductor currents are protected by the same RCD, unwanted tripping may occur. It is therefore necessary to design circuits with such equipment correctly to reduce the likelihood that the RCD will trip unless an unexpected residual current causes it to do so.

In order to test the effectiveness of RCDs after installation, a series of tests may be applied to check that they are within specification and operate satisfactorily. This test sequence will be in addition to proving that the test button is operational. The effectiveness of the test button should be checked after the test sequence.

For each of the tests, readings should be taken on both positive and negative half-cycles and the longer operating time recorded. See Guidance Note 3 and the *On-Site Guide* for more information.

Each type of RCD, e.g. general-purpose RCDs to BS EN 61008, RCBOs to BS EN 61009 or RCD-protected socket-outlets to BS 7288, is initially tested by injecting a current which is 50 per cent of the rated tripping current, i.e. 15 mA; under this condition, the RCD must not operate.

The RCD is then tested by injecting a current which is 100 per cent of the rated tripping current, i.e. 30 mA; under this condition, the RCD must operate within a specified time.

As no test is made between 15 mA and 30 mA, it is reasonable to expect that some RCDs will operate when a residual current within those parameters is flowing. Therefore, designers should be aware of all possibilities when designing circuits. It may be prudent to limit the designed protective conductor current to 0.5 $I_{\Delta n}$ (i.e. 15 mA when the circuit is protected by a 30 mA device) and consider surge currents when all items of equipment are energized simultaneously.

8.2   Water heating

554.1
554.3
554.3.2
554.3.3
It is particularly important that manufacturers' instructions are followed when installing electrode water heaters or boilers, to isolate simultaneously every cable feeding an electrode and to provide overcurrent protection. Similarly, the installation of water heaters having immersed and uninsulated heating elements must follow makers' instructions and be correctly bonded. The metal parts of the water heater (e.g. taps, metal-to-metal joints, covers) must be bonded to the main earthing terminal with a dedicated protective conductor independent of the circuit protective conductor. The heater must be permanently connected through a double-pole switch and not by means of a plug and socket-outlet (see also Guidance Note 2).

8.3   Safety services

560.6.1
560.6.5
Supplies for safety circuits must comply with Chapter 56 and safety sources can be batteries, generating sets or independent incoming feeders. Where incoming independent feeders are intended for use as safety sources, the distribution network operator (DNO) should be consulted to provide an assurance that the two supplies are unlikely to fail concurrently.

560.6.7
A safety source can supply other, non-safety critical circuits where the source has sufficient capacity; in these cases, the circuits are to be configured so that a fault on non-critical circuit does not interrupt the safety circuits.

8.4   Other equipment

559.5.1 Chapter 55 covers requirements for accessories and other equipment, most of which are quite straightforward. Specific requirements are given for typical equipment and accessories. Attention is drawn to Regulation 559.5.1, which allows a choice of accessories to terminate a lighting point.

In domestic installations, use of a device for connecting luminaires (DCL) in accordance with BS EN 61995-1 is recommended, since the interface for this device is harmonized throughout Europe. Additionally, installation of a box at a lighting point, to accommodate the connections, the socket-outlet part of the DCL and the support of a luminaire, has been introduced to overcome the non-regulated but somewhat common and unsafe practice of pushing connections up into a ceiling void. Use of a DCL is intended to permit replacement, cleaning and decoration without the need for access to fixed wiring.

Luminaire supporting couplers (LSCs) or plug and socket-outlet arrangements are alternatives to the harmonized DCL system, each being used to facilitate the removal of luminaires for repair or rearrangement of the lighting layout. Item (v) of Regulation 559.5.1 permits only the use of socket-outlets complying with BS 1363-2, BS 546 or BS EN 60309-2.

The installation of DCLs, LSCs or other socket-outlet systems allows an installation to be completed, tested and energized before luminaires are selected and installed. Luminaires can then be fitted (perhaps after decorating) without interference with the installation and consequent retesting.

Many modern lighting installations are controlled by sophisticated management systems with automatic sensors and software control.

There may not yet be a suitable British Standard for the hardware of such systems but the wiring and the selection and erection of the equipment and luminaire connections must still comply with BS 7671.

511.1 Accessories must comply with the relevant British Standards and care must be exercised
512.2 to ensure damage does not occur during installation, operation or maintenance. Accessories must be of a type and IP rating appropriate to their location and should not be installed in locations where they may suffer impact, wet or dampness or corrosion unless they are of a design specifically for the application.

559.5.1.205 B15 and B22 bayonet lampholders must comply with BS EN 61184, and Regulation 559.5.1.205 requires a T2 temperature rating. This can easily be overlooked by designers and installers. Edison screw lampholders should comply with BS EN 60238:2004, which details temperature ratings.

555.1 The installation of autotransformers and step-up transformers must comply with the requirements of Regulation 555.1. Step-up autotransformers are not permitted in an IT system. Multipole linked switches must be used so that all live conductors, including the neutral if any, can be simultaneously disconnected from the supply (see also Guidance Note 2).

8.5   Luminaires

8.5.1   Scope

559.1   Section 559 applies to the selection and erection of luminaires intended to be part of the fixed lighting installation. The requirements do not apply to high voltage signs such as neon tubes and luminous discharge tube installations exceeding 1 kV, or to temporary festoon lighting. The requirements for high voltage signs are given in the BS EN 50107 series.

8.5.2   General

559.3.1   Luminaires should be manufactured and tested to the relevant standard, i.e. BS EN 60598, however, some cheaper or imported types may not be in compliance with the standard. The installer should be wary of the construction and heat dissipation of these products, and all luminaires should be installed strictly in accordance with the manufacturer's instructions.

559.3.2   Where extra-low voltage luminaires are installed in series, i.e. without low/extra-low voltage transformers, the installation must be treated as a low voltage installation.

8.5.3   Protection against fire

Table 55.2   Luminaires are usually installed on non-combustible surfaces, defined in BS EN 60598-1 as 'material incapable of supporting combustion' and taken to include metal, plaster and concrete. However, sometimes it may be necessary to mount luminaires on materials that may be considered to be flammable (e.g. wood and wood-based materials more than 2 mm thick) and in such cases a classified luminaire to BS EN 60598 marked with the following symbol should be selected:

Luminaires which are not suitable for mounting on normally flammable surfaces are marked as follows:

The following marking indicates the luminaire is suitable for mounting on or in normally flammable material and also has been specifically designed to be covered with insulating material:

Where ceilings or walls provide segregation between fire compartments, the fire rating of any ceiling or wall which is penetrated to install recessed luminaires must be maintained after the installation of the luminaires and the designer should specify

8.6.4 Protection and identification of cables

The definitions in Regulation 1(5) of the ESQC Regulations are intended to include cables supplying street furniture within the scope of the statutory instrument. Such cables are included within the term 'network' and the persons owning or operating such cables are 'distributors' as defined. The implications of this are as follows:

▶ In accordance with Regulation 14, the cables must be buried at sufficient depth to prevent danger and they must be protected or marked.
▶ In accordance with Regulation 15(2), the distributor must maintain maps of the cables.

The DTI guidance on the ESQC Regulations (publication reference URN 02/1544) gives advice on the methods that dutyholders should employ to demonstrate compliance with Regulation 14(3) and thereby reduce the risk of injury to contractors or members of the public. Listed in order of preference the methods are:

(a) cable installed in a duct with marker tape above
(b) cable installed in a duct only
(c) cable laid direct and covered with protective tiles
(d) cable laid direct and covered with marker tape
(e) some other method of mark or indication.

In consideration of the methods by which cables should be marked or protected, dutyholders should make allowance for the environment in which the cables are installed and the risks to those who may need to expose and work on or near the cables in future.

The National Joint Utilities Group (NJUG) have agreed colours for ducts, pipes, cables and marker/warning tapes when laid in the public highway, see Tables 5.2 and 5.3 in Chapter 5 of this Guidance Note.

8.6.5 External influences

512.2
714.512.2.1
Highway equipment suffers from many external influences, including condensation, corrosion, vibration, vandalism, etc., and the designer should take account of such factors in the selection of equipment. Where lamp controlgear is present, the heat generated by the equipment coupled with adequate ventilation will generally be sufficient to combat the effects of condensation and corrosion. However, in other situations the provision of a low wattage heater may be necessary to limit the effects of condensation and extend useful life. Luminaires and other electrical equipment should be secured in a manner to withstand vibration caused by wind and vehicles. All enclosures should be provided with access doors that can only be opened with a tool and are resistant to vandalism.

714.512.2.105
Equipment is required to have a degree of protection against ingress of solid objects and water after erection of at least IP33.

8.6.6 Bus shelters, etc. and decorative lighting

714.411.3.3
Lighting in bus shelters, telephone kiosks and similar must be provided with 30 mA RCD protection.

It is recommended that any decorative lighting (such as Christmas tree lights) within reach of the general public be supplied by SELV. Such equipment is readily available and suitable for indoor and outdoor use.

8.6.7 References

▶ BS 5489 *Code of practice for road lighting.*
▶ NJUG Publication 7 *Recommended positioning of utilities mains and plant for new works.*
▶ Electricity Association Engineering Recommendation G39/2 *Model code of practice covering electrical safety in the planning, installation, commissioning and maintenance of public lighting and other street furniture.*
▶ Institution of Lighting Engineers *Code of practice for electrical safety in public lighting operations.*

8.7 Extra-low voltage lighting installations

8.7.1 Protection against overcurrent

Chap 43   SELV circuits must be protected against overcurrent either by a common protective device or by a protective device for each SELV circuit, in accordance with the requirements of Chapter 43.

8.7.2 Isolation and switching

Where transformers are operated in parallel, the primary circuit must be permanently connected to a common isolating device (see Figure 8.7). The transformers must also have identical electrical characteristics.

▼ **Figure 8.7**   SELV lighting installation with parallel-connected transformers and a common output circuit

Table 51   **Notes to Figure 8.7:**

Table 51   The SELV conductors can be wired using any of the following colour combinations:
▶ two browns
▶ brown and blue
▶ any combination from brown, black, red, orange, yellow, violet, grey, white, pink or turquoise
▶ any colour overmarked with brown or L at terminations with the exception of single-core green-and-yellow, which cannot be used.

8.7.3 Protection against fire

421.1.2 Extra-low voltage lamps and their transformers generate significant heat; fire risks arise from the installation of lamps close to floorboards or joists and the installation of transformers in high ambient temperatures and with limited ventilation. Also, the installer needs to ensure that thermal insulation materials do not interfere with the heat dissipation from the equipment.

8.7.4 Nature of processed or stored materials

The manufacturer's installation instructions must be followed, particularly those relating to mounting on flammable surfaces.

8.7.5 Fire risk of transformers and convertors

715.414 Transformers should be either:

715.422.106 ▶ protected on the primary side by the protective device required for protection against fire, or

715.414 ▶ short-circuit proof (both inherently and non-inherently).

715.414 Electronic convertors supplying filament lamps should comply with BS EN 61347-2-2 and electronic convertors supplying LEDs should comply with BS EN 61347-2-12.

Note: It is recommended that convertors marked with the following symbol are used:

8.7.6 Fire risk from short-circuiting of uninsulated conductors

715.422.107 If both circuit conductors are uninsulated, then either:

▶ they must be provided with a protective device complying with the requirements below, or
▶ the system must comply with BS EN 60598-2-23Particular requirements. ELV lighting systems for filament lamps, or
▶ they must be supplied from a transformer or converter, the power of which does not exceed 200 VA.

715.422.107.2 Where the first option is adopted to provide protection against the risk of fire, the device must:

▶ continuously monitor the power demand of the luminaires;
▶ automatically disconnect the supply circuit within 0.3 s in the event of a short-circuit or failure that causes a power increase of more than 60 W;
▶ automatically disconnect while the supply circuit is operating with reduced power (for example, by gating control or a regulating process or a lamp failure) if there is a failure that causes a power increase of more than 60 W;
▶ automatically disconnect during switching of the supply circuit if there is a failure that causes a power increase of more than 60 W; and
▶ be fail-safe.

8.7.7 Bare conductors

715.521.106   If the nominal voltage does not exceed 25 V a.c. or 60 V d.c., bare conductors may be used provided that the extra-low voltage lighting installation complies with the following requirements:

▶ the lighting installation is designed and installed or enclosed in such a way that the risk of a short-circuit is reduced to a minimum, and
▶ the conductors used have a cross-sectional area of at least 4 mm², for mechanical reasons, and
▶ the conductors or wires are not placed directly on combustible material.

For suspended bare conductors, at least one conductor and its terminals must be insulated for that part of the circuit between the transformer and the protective device, to prevent the occurrence of short-circuit.

8.7.8 Suspended systems

715.521.107   If the fixing accessory is intended to support a pendant luminaire, the accessory should be capable of carrying a mass of not less than 5 kg. If the mass of the luminaire is greater than 5 kg, the installer should ensure that the fixing means is capable of supporting its mass. The installation instructions of the manufacturer should be followed.

Termination and connection of conductors should be by screw terminals or screwless clamping devices complying with BS EN 60998-2-1 or BS EN 60998-2-2.

Insulation piercing connectors and termination wires that rely on counterweights hung over suspended conductors to maintain the electrical connection must not be used.

The suspended system should be fixed to walls or ceilings by insulated distance cleats and must be continuously accessible throughout the route.

8.7.9 Cross-sectional area of conductors

715.524   The minimum cross-sectional area of the extra-low voltage conductors should be:

▶ 1.5 mm² copper, but in the case of flexible cables with a maximum length of 3 m a cross-sectional area of 1 mm² copper may be used
▶ 4 mm² copper in the case of suspended flexible cables or insulated conductors, for mechanical reasons
▶ 4 mm² copper in the case of composite cables consisting of braided tinned copper outer sheath, having a material of high tensile strength inner core.

8.8   Selection and erection in relation to operation and maintainability

The Electricity at Work Regulations 1989 specifically place a duty on installation or equipment owners or managers (dutyholders) to ensure the installed equipment is utilized safely and persons are protected from danger. In the workplace, ordinary persons are not permitted to access live systems and equipment or carry out electrical work.

Where functions are not immediately obvious, operating instructions should be displayed adjacent to that item. All other equipment should be located behind lockable doors or accessible only by the use of a tool. The building operator should be provided with full instructions as to the operation and maintenance of the installation, including mechanical operation.

341.1 BS 7671 requires an assessment of expected maintenance to be made as part of the design process. Selection of equipment should be relevant to the maintenance (or lack of) it may receive during its installed and operating life.

For all items that provide physical protection which may need removal (including lids of cable boxes, trunking, etc.) and replacement, the procedure should be as simple as possible. It should be possible to be carried out by one person without assistance. Where frequent access is likely, lids, etc., should be hinged. It should also be possible to close all lids and covers without putting pressure on enclosed cables and equipment.

132.12 513.1 529.3 Electrical distribution equipment and switchgear should be located so that all components may be operated and maintained safely. The Electricity at Work Regulations 1989 (Regulation 15) specifically require adequate working space, access and lighting to be provided for all electrical equipment, where working on or near it may give rise to danger.

Appendix A

Cable capacities of conduit and trunking

A1 General

This appendix describes a method based on practical work and experimentation which can be used to determine the size of conduit or trunking necessary to accommodate cables of the same size or differing sizes, and provides a means of compliance with the requirements of Chapter 52 of BS 7671.

The method employs a 'unit system', each cable size being allocated a factor. The sum of all factors for the cables intended to be run in the same enclosure is compared against the factors given for conduit, ducting or trunking, as appropriate, in order to determine the size of the conduit or trunking necessary to accommodate those cables.

It has been found necessary, for conduit, to distinguish between:

▶ case 1 – straight runs not exceeding 3 m in length, and
▶ case 2 – straight runs exceeding 3 m, or runs of any length incorporating bends or sets.

The term 'bend' signifies a British Standard 90 ° bend and one double set is equivalent to one bend.

For case 1, each conduit size is represented by only one factor. For case 2, each conduit size has a variable factor which is dependent on the length of run and the number of bends or sets. For a particular size of cable the factor allocated to it for case 1 is not the same as for case 2.

For trunking, each size of cable has been allocated a factor, as has been each size of trunking.

A number of variables affect any attempt to arrive at a standard method of assessing the capacity of conduit or trunking. Some of these are:

(a) reasonable care in installation
(b) acceptable use of the space available
(c) tolerance in cable sizes
(d) tolerance in conduit and trunking.

The following tables can only give guidance on the maximum number of cables which should be drawn in. The sizes should ensure an easy pull with low risk of damage to the cables.

Only the ease of drawing-in is taken into account. The electrical effects of grouping

▶ **Table A4** Cable factors for runs incorporating bends and long straight runs

Length of run (m)	Straight				One bend				Two bends				Three bends				Four bends			
Conduit diameter (mm)	16	20	25	32	16	20	25	32	16	20	25	32	16	20	25	32	16	20	25	32
1		Covered by Tables A1 and A2			188	303	543	947	177	286	514	900	158	256	463	818	130	213	388	692
1.5					182	294	528	923	167	270	487	857	143	233	422	750	111	182	333	600
2					177	286	514	900	158	256	463	818	130	213	388	692	97	159	292	529
2.5					171	278	500	878	150	244	442	783	120	196	358	643	86	141	260	474
3					167	270	487	857	143	233	422	750	111	182	333	600				
3.5	179	290	521	911	162	263	475	837	136	222	404	720	103	169	311	563				
4	177	286	514	900	158	256	463	818	130	213	388	692	97	159	292	529				
4.5	174	282	507	889	154	250	452	800	125	204	373	667	91	149	275	500				
5	171	278	500	878	150	244	442	783	120	196	358	643	86	141	260	474				
6	167	270	487	857	143	233	422	750	111	182	333	600								
7	162	263	475	837	136	222	404	720	103	169	311	563								
8	158	256	463	818	130	213	388	692	97	159	292	529								
9	154	250	452	800	125	204	373	667	91	149	275	500								
10	150	244	442	783	120	196	358	643	86	141	260	474								

Additional factors:

▲ For 38 mm diameter use 1.4 × (32 mm factor)
▲ For 50 mm diameter use 2.6 × (32 mm factor)
▲ For 63 mm diameter use 4.2 × (32 mm factor)

A4 Single-core thermoplastic (PVC) insulated cables in trunking

For each cable it is intended to use, obtain the appropriate factor from Table A5.

Add the cable factors together and compare the total with the factors for trunking in Table A6.

The minimum size of trunking is that size having a factor equal to or greater than the sum of the cable factors.

▼ **Table A5** Cable factors for trunking

Type of conductor	Conductor cross-sectional area (mm^2)	Thermoplastic (PVC) BS 6004 Table 1 Cable factor	Thermosetting BS 7211 Table 3 Cable factor
Solid	1.5	8.0	8.6
	2.5	11.9	11.9
Stranded	1.5	8.6	9.6
	2.5	12.6	13.9
	4	16.6	18.1
	6	21.2	22.9
	10	35.3	36.3
	16	47.8	50.3
	25	73.9	75.4
	35	93.3	95.1
	50	128.7	132.8
	70	167.4	176.7
	95	229.7	227.0
	120	277.6	283.5
	150	343.1	346.4
	185	426.4	433.7
	240	555.7	551.6

Notes:

(a) Cable factors are the cable cross-sectional area using the BS upper limit mean overall diameter

(b) The provision of spare space is advisable; however, any circuits added at a later date must take into account grouping (see Appendix 4 of BS 7671 for further details)

(c) Where thermosetting insulated conductors designed to operate at 90 °C (BS 5467 or BS 7211 etc.) are installed together with thermoplastic (PVC) insulated conductors designed to operate at 70 °C, it must be ascertained that the thermoplastic (PVC) insulated conductors will not be damaged. (Regulation 523.5.)

Appendix B

Degrees of protection provided by enclosures

B1   IP code for ingress protection

B1.1   General

The requirements of the IP code are given in BS EN 60529. This is a standard used to form the basic requirements of electrical equipment standards and the construction for IP ratings of a specific type of equipment is given in the BS for that type of equipment.

The degree of protection provided by an enclosure is indicated by two numerals followed by an optional additional letter and/or optional supplementary letter(s) as shown in Figure B1.

▼ **Figure B1**   IP code format

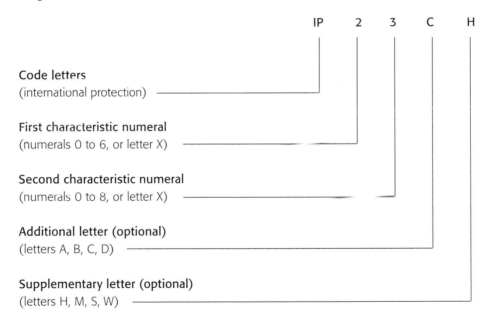

Where a characteristic numeral is not required to be specified, it can be replaced by the letter 'X' ('XX' if both numbers are omitted).

The additional letter and/or supplementary letter(s) may be omitted without replacement. Where more than one supplementary letter is used, the alphabetic sequence is used.

B1.4   Product marking

The requirements for marking a product are specified in the relevant product standard. An example could be:

IP23CS

An enclosure with this designation (IP code):

2   protects persons against access to hazardous parts with a finger and protects the equipment inside the enclosure against ingress of solid foreign objects having a diameter of 12.5 mm and greater.

3   protects the equipment inside the enclosure against the harmful effects of water sprayed against the enclosure at an angle of up to 60 ° either side of the vertical.

C   protects persons handling tools having a diameter of 2.5 mm and greater and a length not exceeding 100 mm against access to hazardous parts (the tool may penetrate the enclosure up to its full length).

S   is tested for protection against harmful effects due to the ingress of water when all the parts of the equipment are stationary.

If an enclosure provides different degrees of protection for different intended mounting arrangements, the relevant degrees of protection must be indicated by the manufacturer in the instructions relating to the respective mounting arrangements.

B1.5   Drip-proof and splashproof

Certain electrical equipment has been graded against water ingress by the identifications 'Drip Proof', etc. This labelling has been superseded by the IP coding but is given in Figure B2 with an equivalent IP rating for comparison. Comparisons are not exact and the designer or installer must satisfy themselves that any equipment is suitable for the location of use.

▼ **Figure B2**   Superseded water ingress labels and their equivalent IP ratings

Drip-proof		IPX2
Rain proof		IPX3
Splashproof		IPX4
Jet proof		IPX5
Protected against immersion in water		IPX7

B2   IK code for impact protection

B2.1   General

BS EN 62262:2002 *Degrees of protection provided by enclosures for electrical equipment against external mechanical impacts (IK code)* specifies a system for classifying the degrees of protection provided by enclosures against mechanical impact.

The standard describes only the general requirements and designations for the system – the application of the system to a specific enclosure type will be covered by the British Standard applicable to that equipment or enclosure. An enclosure is defined as a part providing protection of equipment against certain external influences and protection against contact. This may be considered to include conduit, trunking, etc.

In general, the degree of protection will apply to a complete enclosure. If parts of an enclosure have different degrees of protection, they must be separately identified. The coding is separate from the IP rating and will be marked separately in the following way:

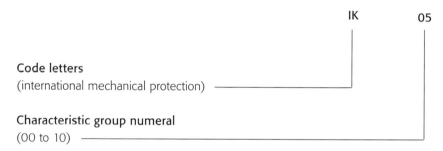

Code letters
(international mechanical protection)

Characteristic group numeral
(00 to 10)

Each characteristic group numeral represents an impact energy value as shown below:

IK Code	IK00	IK01	IK02	IK03	IK04	IK05	IK06	IK07	IK08	IK09	IK10
Impact energy in joules	*	0.15	0.2	0.35	0.5	0.7	1	2	5	10	20

* No protection specified

Where higher impact energy is required the value of 50 joules is recommended.

does not exceed the value given in Table C1.

311.1
132.3 An assessment of the loading must be made for the design of an installation, in accordance with Chapter 31 of BS 7671, and adequate circuits provided. (See also Regulation 132.3.)

For household installations, a single 30 A or 32 A ring final circuit may serve a floor area of up to 100 m² but consideration should be given to the loading in kitchens which may require a separate circuit. Socket-outlets for washing machines, tumble driers and dishwashers should be located so as to provide reasonable sharing of the load in each leg of the ring, or consideration should be given to a 4 mm² radial circuit. 2.5 mm² radial circuits may also adequately serve areas of a dwelling other than kitchens. For other types of premises, final circuits complying with Table C1 may be installed where, owing to diversity, the maximum demand of current-using equipment to be connected is estimated not to exceed the corresponding ratings of the circuit cables.

553.1.7 The number of socket-outlets should be such as to ensure compliance with Regulation 553.1.7, each socket-outlet of a twin or multiple socket-outlet unit being regarded as one socket-outlet.

Diversity between socket-outlets and permanently connected equipment has already been taken into account in Table C1 and no further diversity may be applied.

▼ **Table C1** Typical final circuits using BS 1363 socket-outlets in household installations

		Minimum conductor cross-sectional area* (mm²)			
Type of circuit	Overcurrent protective device rating (A)	Copper conductor thermoplastic or thermosetting insulated cables	Copper conductor mineral insulated cables	Maximum floor area served‡ (m²)	
1	2	3	4	5	6
A1	Ring	30 or 32	2.5	1.5†	100
A2	Radial	30 or 32	4	2.5	75
A3	Radial	20	2.5	1.5†	50

*   The tabulated values of conductor size may need to be increased where more than two circuits are grouped together but may be reduced for fused spurs. Table C1 applies to thermoplastic or thermosetting insulated cables of the types identified in Table 4A2 of Appendix 4 of BS 7671 in installation method C and method B and installed in accordance with those installation methods, and mineral insulated cable (MICC) as identified in Table 4G1A installed in accordance with reference methods C, E, F and G of Table 4A2. It does not apply to 70 °C thermoplastic (PVC) insulated and sheathed flat twin cable with protective conductor installed in accordance with method A of Table 4D5.

†   1.5 mm² two-core MICC installed for a ring final circuit with a protective device rated at not more than 32 A or a radial circuit with a protective device rated at not more than 20 A may be taken as having a conductor not operating at a temperature above 70 °C and may be used to comply with the relevant requirements of Regulation 512.1.2. 1.5 mm² four-core MICC will not comply with Regulation 433.1.204 at 70 °C.

‡   The socket-outlets are to be as evenly distributed between circuits as is reasonable.

Where two or more ring final circuits are installed, the socket-outlets and permanently connected equipment to be served should be reasonably distributed among the circuits.

C2.2 Circuit protection

Table C1 is applicable for circuits protected by fuses to BS 3036 and BS 88 and circuit-breakers, types B and C to BS EN 60898 or BS EN 61009-1 and BS EN 60947-2 and types 1, 2 and 3 to BS 3871.

C2.3 Conductor size

The minimum size of conductor in the circuit and in non-fused spurs is given in Table C1. However, if the cables of more than two circuits are bunched together or the ambient temperature exceeds 30 °C, the size of conductor is increased and is determined by applying the appropriate rating factors from Appendix 4 of BS 7671 such that the size then corresponds to a current-carrying capacity not less than:

433.1.204
- ▶ 20 A for ring circuit A1
- ▶ 30 A or 32 A for radial circuit A2 (i.e. the rating of the overcurrent protective device)
- ▶ 20 A for radial circuit A3 (i.e. the rating of the overcurrent protective device).

The conductor size for a fused spur is determined from the total current demand served by that spur, which is limited to a maximum of 13 A.

Where a fused spur serves socket-outlets the minimum conductor size is:

- ▶ 1.5 mm^2 for thermoplastic (PVC) or thermosetting insulated cables, with copper conductors
- ▶ 1 mm^2 for mineral insulated cables, with copper conductors.

433.1.204
433.1.1
For a 30 A or 32 A ring final circuit with a protective device in accordance with the requirements of Regulation 433.1.204, supplying 13 A socket-outlets to BS 1363, a conductor with a minimum csa of 2.5 mm^2 is deemed to comply with the requirements of Regulation 433.1.1 if the current-carrying capacity (I_z) of the cable is not less than 20 A and if, under the intended conditions of use, the load current in any part of the ring is unlikely to exceed for long periods the current-carrying capacity (I_z) of the cable.

Electrical accessories to BS 1363, etc., are designed for a maximum operating temperature at their terminals and also for use with thermoplastic (PVC) insulated conductors operating at a maximum conductor temperature of 70 °C.

Table 52.1
Thermosetting cables can operate with conductor temperatures up to 90 °C and some types of mineral insulated cable can operate with conductor temperatures up to 105 °C.

512.1.5
Such temperatures can damage accessories (see Regulation 512.1.5 of BS 7671). Consequently, when used, such cables must be derated to comply with this requirement (see note 1 of Table 4E2A, etc. of Appendix 4 of BS 7671).

C2.4 Spurs

The total number of fused spurs is unlimited but the number of non-fused spurs is not to exceed the total number of socket-outlets and items of stationary equipment connected directly in the circuit.

A non-fused spur feeds only one single or one twin or multiple socket-outlet or one permanently connected item of electrical equipment. Such a spur is connected to a circuit at the terminals of socket-outlets or at junction boxes or at the origin of the circuit in the distribution board.

A fused spur is connected to the circuit through a fused connection unit, the rating of the fuse in the unit not exceeding that of the cable forming the spur and, in any event, not exceeding 13 A. The number of socket-outlets which may be supplied by a fused spur is unlimited.

C2.5   Permanently connected equipment

537.5.1.3 Permanently connected equipment, such as built-in washing machines, electric ovens, etc., is often connected to the supply by means of a plug and socket-outlet located behind the appliance. A separate control switch need not be installed provided that:

- ▶ a functional switch is incorporated in the appliance, and

    - the appliance is locally protected by a fuse to BS 1362, of a rating not exceeding 13 A, or
    - the circuit is protected by a circuit-breaker of a rating not exceeding 16 A.

C3   Radial final circuits using 16 A socket-outlets complying with BS EN 60309-2 (BS 4343)

C3.1   General

Where a radial final circuit feeds equipment the maximum demand of which, having allowed for diversity, is known or estimated not to exceed the rating of the overcurrent protective device and, in any event, does not exceed 20 A, the number of socket-outlets is unlimited.

C3.2   Circuit protection

The overcurrent protective device has a rating not exceeding 20 A.

C3.3   Conductor size

The size of conductor is determined from Appendix 4 of BS 7671 by applying the appropriate rating factors and is such that it then corresponds to a current-carrying capacity not less than the rating of the overcurrent protective device.

C3.4   Types of socket-outlet

Socket-outlets have a rated current of 16 A and are of the type appropriate to the number of phases, circuit voltage and earthing arrangement. Socket-outlets incorporating pilot contacts are not included.

C4   Cooker final circuits in household premises

The final circuit supplies a control switch complying with BS EN 60669-1 (BS 3676) or a cooker control unit (with a socket-outlet incorporated) complying with BS 4177. Assessment of the risks associated with the installation of a cooker control unit should always be made. A socket-outlet near a cooker could invite accidents if inappropriately located, by allowing cables or equipment near hot surfaces. A cooker control unit with a socket-outlet should be installed so as to minimize the risk of such accidents; if this is not possible it is perhaps better that cooker control units should not incorporate a socket-outlet. A modern kitchen should have an adequate number of socket-outlets.

The rating of a cooking appliance circuit is determined by the assessment of the current demand of the cooking appliance(s), and control unit socket-outlet if any, in accordance with Table H1 of Appendix H. A 30/32 A circuit is suitable for most household cookers but a 40/45 A circuit may be necessary for larger cookers.

537.5.1.3
536.1
A circuit of rating exceeding 15 A but not exceeding 50 A may supply two or more cooking appliances where these are installed in one room. One switch may be used to control all the appliances. The control switch should be placed in a readily accessible position. Attention is drawn to the need to afford discriminative operation of protective devices as stated in Regulation 536.1.

C5 Water heating and electric shower final circuits in household premises

The final circuit shall supply the full load of the electric shower. No diversity is allowable on the final circuit.

Immersion heaters fitted to storage vessels in excess of 15 litres capacity, or permanently connected heating appliances forming part of a comprehensive electric space heating installation, should be supplied by their own separate circuits.

554.3.3
Immersion heaters should not be connected by a plug and socket-outlet but by a switched connection unit complying with BS 1363-4.

C6 Provision of socket-outlets

There is no statutory requirement for the minimum provision of socket-outlets but the number of socket-outlets needed in a dwelling has risen significantly over the years and can be expected to continue to rise as more electrical goods become available and people's working practices change, with working from home now quite common. Table C2 makes proposals for the provision of socket-outlets in modern dwellings.

C6.1 Number of socket-outlets

553.1.7
To satisfy Regulation 553.1.7 sufficient socket-outlets must be provided and located to allow the use of mobile equipment with the length of flexible cable normally fitted. Long extension leads, trailing sockets and extending flexible cables which are incorrectly or inappropriately located can create tripping hazards and can damage and cause wear and overheating of the flex when flexible cables are placed under carpets, etc. In order, so far as is possible, to make the use of multiway adaptors or extension leads unnecessary, an adequate number of fixed socket-outlets should be installed as listed in Table C2. Multiway adaptors, long extension leads, trailing socket-outlets and extending flexible cables should never be used unattended and should always be visible and accessible throughout their length.

Unfortunately, no matter how good the overall layout some socket-outlets will inevitably become inaccessible behind furniture when the dwelling is furnished and the designer should consider this in the initial provisions. Also, the continued future growth in appliances and equipment should be expected and so what may seem initially an over-generous provision ultimately may not be.

▼ **Table C2** Minimum number of twin socket-outlets to be provided in homes

Room type	Smaller rooms (up to 12 m^2)	Medium rooms (12–25 m^2)	Larger rooms (more than 25 m^2)
Main living room (note 4)	4	6	8
Dining room	3	4	5
Single bedroom (note 3)	2	3	4
Double bedroom (note 3)	3	4	5
Bedsitting room (note 6)	4	5	6
Study	4	5	6
Utility room	3	4	5
Kitchen (note 1)	6	8	10
Garage (note 2)	2	3	4
Conservatory	3	4	5
Hallway	1	2	3
Loft	1	2	3
Locations containing a bath or shower	note 5		

(Table by kind permission of the ECA, SELECT and BEAMA.)

Notes:

1 **KITCHEN** – If a socket-outlet is provided in the cooker control unit, this should not be included in the 6 recommended in the table above.

   Appliances built into kitchen furniture (integrated appliances) should be connected to a socket-outlet or switch fused connection unit that is accessible when the appliance is in place and in normal use. Alternatively, where an appliance is supplied from a socket-outlet or a connection unit, these should be controlled by an accessible double-pole switch or switched fused connection unit.

   It is recommended that wall mounted socket-outlets above a work surface are spaced at not more than 1 m intervals along the surface.

2 **GARAGE** – The number of socket-outlets specified allows for the use of a battery charger, tools, portable light and garden appliances.

3 **BEDROOM** – It is envisaged that this room will be used in different ways in different households. It may be used simply as a child's bedroom requiring socket-outlets for table lamps, an electric blanket and an electric heater only; or it may serve as a teenager's bedroom and living room combined, where friends are entertained. In this case, socket-outlets may be needed for computers (printers, scanners, speakers, etc.), games consoles, MP3/4 players, mobile phone chargers, DVD players, digital receivers, home entertainment systems (amplifier, CD player), hairdryer, television and radio, in addition to lamps, an electric blanket and electric heater.

4 **HOME ENTERTAINMENT** – In addition to the number of socket-outlets shown in the table it is recommended that at least two further double socket-outlets are installed in home entertainment areas.

701.512.3 **5 LOCATIONS CONTAINING A BATH OR SHOWER** – Except for SELV socket-outlets complying with Section 414 and shaver supply units complying with BS EN 61558-2-5, socket-outlets are prohibited within a distance of 3 m horizontally from the boundary of zone 1.

6 BEDSITTING ROOM – Rooms specifically designed or envisaged to be used as student bedsitting rooms should be provided with additional socket-outlets which may be needed since persons using these rooms will often introduce other portable appliances in addition to items already mentioned in Note 3. In such situations a lack of sufficient socket-outlets is an additional danger and therefore the minimum number of twin outlets should be increased to four.

C6.2 Sinks and electrical accessories

How close are electrical accessories, such as socket-outlets, permitted to be to a sink? This question is commonly asked, not only by members of the public but also by electrical contractors and designers concerning both domestic and commercial premises.

Some are under the impression that a minimum distance is given in BS 7671 within which electrical accessories must not be installed. This is not true. It is impractical to prescribe such a distance, given the fact that many kitchens are very small. It is almost inevitable therefore, in many situations, that socket-outlets will be installed which are simultaneously accessible to sink tops.

132.5.1
512.2.1
The type of accessory installed must be of a design appropriate to its location as required by Regulations 132.5.1 and 512.2.1. Where BS 1363 accessories are used, they should be installed as far from the sink as practicable. As a general recommendation, BS 1363 accessories should be mounted above sink height at a horizontal distance of not less than 300 mm from the edge of the sink top.

C7 Mounting heights of accessories

553.1.6
Accessories must be mounted at a height that will make them accessible but minimize the risk of mechanical damage by their being struck by floor cleaning equipment or the movement of furniture, and allow a comfortable bending radius for larger or stiff flexes. The Building Regulations 2000 Approved Document M (for England and Wales), *Access and facilities for disabled people*, contains appropriate mounting heights between 450 mm and 1200 mm for electrical equipment in dwellings (see Figure C1). The Scottish Building Standards specify similar requirements.

▼ **Figure C1**   Height of switches, sockets, etc. (see Approved Document M, section 8)

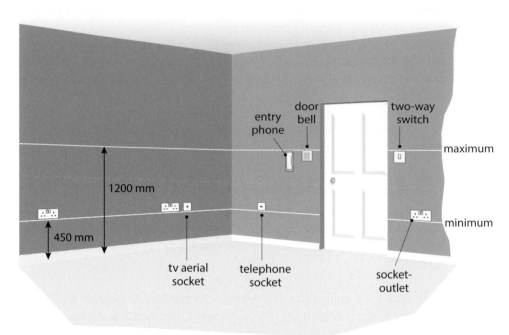

C7.1   BS 8300:2009+A1:2010 Design of buildings and their approaches to meet the needs of disabled people – Code of practice

BS 8300, which applies to all types of buildings, contains a number of recommendations which could affect the design of an electrical installation. The requirements concerning the location of outlets, switches, controls and meters are shown in the extract from the standard given in Figure C2.

It should be noted that this document is not mandatory unless directly referred to in the specification, or the specification contains a general requirement to the effect that the electrical installation shall comply with all relevant standards and codes. In case of doubt, the exact requirements of the client must be ascertained prior to undertaking the project.

▼ **Figure C2**   Extract from BS 8300:2009

10.5.2   Location of outlets, switches, controls and meters

All outlets, switches and controls, including two-way switching, should be positioned consistently in relation to doorways and corners within a building and in a logical sequence to suit passage through the building.

Note 1: By using vertical strips as switches, instead of single height switches, users can operate the switch at whichever height is convenient for them.

Preferably, light switches should align horizontally with door handles for ease of location when entering a room.

Electrical socket-outlets, telephone points and TV sockets should be located at least 400 mm but not more than 1000 mm above the floor. Socket-outlets whose plugs are likely to be removed and replaced frequently should be located at the top of the range (see figure below).

Note 2: Switches close to the floor or skirting are difficult and dangerous because they require users to stoop or kneel to operate them. The higher the socket-outlet, the easier it is to push in or pull out the plug.

Switches for permanently wired appliances (e.g. fused spurs or reset switches for alarm calls) should be mounted within the range between 750 mm and 1200 mm (see figure below).

Meters should be mounted between 1200 mm and 1400 mm from the floor so that the readings can be viewed by a person standing or sitting. Pre-pay meters should be accessible but protected so children cannot tamper with them.

All switches and controls that require precise hand movement/dexterity, e.g. for heating installations, ventilation, etc., should be in a zone 750 mm to 1000 mm from the floor (see figure below) so that wheelchair users and those standing can operate them.

The maximum height of simple push button controls, including isolator switches and circuit-breakers, that require limited dexterity should be 1200 mm (see figure below).

Outlets, switches and controls should be at least 350 mm from room corners.

C

Guidance Note 1: Selection & Erection

Appendix D

Limitation of earth fault loop impedance for compliance with Regulation 543.1.1

Regulation 543.1.1 indicates that the cross-sectional area of a protective conductor, other than a protective bonding conductor, shall be:

1 calculated in accordance with Regulation 543.1.3, or
2 determined in accordance with Regulation 543.1.4.

In some cases the type of wiring system selected for use will determine which of the two methods can be followed. For instance, the widely used flat twin and flat three-core thermoplastic (PVC) insulated and thermoplastic (PVC) sheathed cables with protective conductors (cables to Table 5 of BS 6004) do not comply with Table 54.7 of Regulation 543.1.4 (other than the 1 mm^2 size) and therefore method 1 should be used.

Appx 3 Where method 1 is used, in order to apply the formula given in Regulation 543.1.3 it is essential that the time/current characteristic of the overcurrent protective device in the circuit concerned is available. A selection of such characteristics for fuses and circuit-breakers is given in Appendix 3 of BS 7671. For other types of device the advice of the manufacturer has to be sought. The time/current characteristics given in Appendix 3 indicate the values of I_a against various disconnection times for the devices tabulated and give coordinates for fixed times. For circuit-breakers, the curves indicate the maximum operating current, I_a, of a particular type of circuit-breaker and RCBO (see Table 5.3 section 3.5.3 of this Guidance Note for CB type details). For HRC fuses, the curves given in Appendix 3 are based on the maximum current for time values in the applicable fuse standards.

543.2.1 Assuming that the size and type of cables to be used have already been determined from consideration of other aspects, such as mechanical protection and installation requirements, together with the magnitude of the design current of the circuit and the limitation of voltage drop under normal load conditions, the first stage is to calculate the earth fault loop impedance, Z_s. If the cable does not incorporate a suitable protective conductor, that conductor has to be chosen separately. Regulation 543.2.1 details what may be used as a protective conductor.

For cables having conductors of cross-sectional area not exceeding 35 mm^2, their inductive reactance can be ignored so that where these cables are used in radial circuits, the earth fault loop impedance Z_s is given by:

$$Z_s = Z_e + (R_1 + R_2) \ \Omega \qquad\qquad (1)$$

Where the adiabatic calculation of S for compliance with Regulation 543.1.3 is carried out, the value of t in equation (4) is 0.1.

For circuits protected by circuit-breakers, compliance with Table D4 provides compliance with Regulation 543.1.1 where the protective conductors range from 1 mm² to 16 mm² cross-sectional area and the rated current of the circuit-breakers ranges from 5 A to 125 A.

In each table the earth fault loop impedances given correspond to the appropriate disconnection time from a comparison of the time/current characteristic of the device concerned and the equation given in Regulation 543.1.3.

The tabulated values apply only when the nominal voltage to Earth (U_0) is 230 V. The circuit loop impedances have been determined using a value for factor C_{min} of 0.95.

For guidance on conductor resistances, see Appendix E.

The tabulated loop impedances are design figures assuming conductor temperatures appropriate to earth fault conditions, which are as described earlier in this appendix. For testing purposes the loop impedances must be reduced. For example, if testing at an ambient temperature of 10 °C using cables to Table 5 of BS 6004, the maximum permitted test loop impedance is given by:

$$Z_{test} \leq (Z_{FX} - Z_e) \frac{0.96}{1.20} + Z_e$$

(5)

where:

Z_{FX} is the loop impedance given by Table D1, D2, D3 or D4

Z_e is the supply earth loop impedance

1.20   is the multiplier from Table E1

0.96   is the ambient temperature correction factor from Table E3

▼ **Table D1** Maximum earth fault loop impedance (in ohms) when overcurrent protective device is a semi-enclosed fuse to BS 3036

i   0.4 s disconnection (final circuits in TN systems, 32 A or less)

Protective conductor (mm²)	Fuse rating (A)			
	5	15	20	30
1.0	9.1	2.43	1.69	NP
1.5 to 16.0	9.1	2.43	1.69	1.04

ii   5 s disconnection (final circuits exceeding 32 A and distribution circuits in TN systems)

Protective conductor (mm²)	Fuse rating (A)			
	15	20	30	45
1.0	5.08	3.19*	NP	NP
1.5	5.08	3.64	2.37*	NP
2.5	5.08	3.64	2.5	1.37*
4.0 to 16.0	5.08	3.64	2.5	1.5

Notes:

▶ A value for k of 115 from Table 54.3 of BS 7671 is used for this table. This is also suitable for cables to Table 5 of BS 6004.

* Limited by thermal considerations of the protective conductor cross-sectional area, see Regulation 543.1.3.

NP Protective conductor, fuse combination not permitted.

▼ **Table D2** Maximum earth fault loop impedance (in ohms) when overcurrent protective device is a fuse to BS 88-2

i 0.4 s disconnection (final circuits in TN systems, 32 A or less)

Protective conductor (mm²)	Fuse rating (A)				
	6	10	16	20	32
1.0	7.8	4.65	2.43	1.68	0.78*
1.5 to 16.0	7.8	4.65	2.43	1.68	0.99

ii 5 s disconnection (final circuits exceeding 32 A and distribution circuits in TN systems)

Protective conductor (mm²)	Fuse rating (A)						
	6	10	16	20	32	40	50
1.0	12.14	6.8	3.64*	1.98*	0.76*	NP	NP
1.5	12.14	6.8	3.97	2.8	1.11*	NP	NP
2.5	12.14	6.8	3.97	2.8	1.75	0.98*	0.65*
4.0	12.14	6.8	3.97	2.8	1.75	1.29	0.91*
6.0 to 16.0	12.14	6.8	3.97	2.8	1.75	1.29	0.99

Notes:

▶ A value for k of 115 from Table 54.3 of BS 7671 is used for this table. This is also suitable for cables to Table 5 of BS 6004.

* Limited by thermal considerations of the protective conductor cross-sectional area, see Regulation 543.1.3.

NP Protective conductor, fuse combination not permitted.

▼ **Table D3** Maximum earth fault loop impedance (in ohms) when overcurrent protective device is a fuse to BS 88-3 (and existing BS 1361 fuses)

i 0.4 s disconnection (final circuits in TN systems, 32 A or less)

Protective conductor (mm²)	Fuse rating (A)			
	5	15	20	30
1.0	9.9	3.12	1.62	0.91*
1.5 to 16.0	9.9	3.12	1.62	1.09

ii 5 s disconnection (final circuits exceeding 32 A and distribution circuits in TN systems)

Protective conductor (mm²)	Fuse rating (A)				
	5	15	20	30	45
1.0	15.6	4.75	1.98*	0.9*	NP
1.5	15.6	4.75	2.55*	1.37*	NP
2.5	15.6	4.75	2.66	1.74	0.6*
4.0	15.6	4.75	2.66	1.74	0.84*
6.0 to 16.0	15.6	4.75	2.66	1.74	0.91

Notes:

▶ A value for k of 115 from Table 54.3 of BS 7671 is used for this table. This is also suitable for cables to Table 5 of BS 6004.

* Limited by thermal considerations of the protective conductor cross-sectional area, see Regulation 543.1.3.

NP Protective conductor, fuse combination not permitted.

▼ **Table D4i** Maximum earth fault loop impedance (in ohms) when overcurrent protective devices are circuit-breakers to BS EN 60898 or RCBOs to BS EN 61009-1

Circuit-breaker type	Circuit-breaker rating (A)											
	6	10	16	20	25	32	40	50	63	80	100	125
B	7.28	4.37	2.73	2.19	1.75	1.37	1.09	0.87	0.69	0.54	0.44	0.35
C	3.6	2.19	1.37	1.09	0.87	0.68	0.54	0.43	0.34	0.27	0.21	0.17
D	1.8	1.09	0.69	0.54	0.43	0.34	0.27	0.22	0.17	0.13	0.1	0.085

▼ **Table D4ii** Maximum earth fault loop impedance (in ohms) when overcurrent protective devices are magnetic circuit-breakers to (now withdrawn) BS 3871

Circuit-breaker type	Circuit-breaker rating (A)											
	5	6	10	15	16	20	25	30	32	40	45	50	63
1	10.92	9.10	5.46	3.64	3.41	2.72	2.18	1.82	1.71	1.37	1.20	1.09	0.86
2	6.24	5.20	3.13	2.08	1.95	1.56	1.24	1.03	0.97	0.78	0.69	0.63	0.49
3	4.37	3.64	2.18	1.45	1.37	1.09	0.87	0.73	0.68	0.54	0.48	0.44	0.34

Notes:

(a) A value for k of 115 from Table 54.3 of BS 7671 is used. This is suitable for thermoplastic (PVC) insulated and sheathed cables to Table 5 and Table 6 of BS 6004 and for LSOH (low smoke zero halogen) insulated and sheathed cables to Table 7 of BS 7211. The k value is based on both the thermoplastic (PVC) and LSOH cables operating at a maximum operating temperature of 70 °C.

411.4
411.3.3
(b) In TN systems it is preferable for reliable operation for fault protection to be provided by overcurrent devices, including RCBOs operating as overcurrent devices; that is, with loop impedance complying with the table above. RCBOs are then providing fault protection as circuit-breakers and additional protection as RCDs. If, however, the residual current element of an RCBO has been selected to provide fault protection, the tabulated values of maximum earth fault loop impedance do not apply.

D1   Selection of a circuit protective conductor

As previously stated, Regulation 543.1.1 allows the selection of a circuit protective conductor (cpc) by calculation in accordance with Regulation 543.1.3 or by reference to Table 54.7 to select the minimum size, in accordance with Regulation 543.1.4.

Table 54.7 provides a method to establish the minimum size of cpc based on the area of the line conductor of the circuit. In many cases the material and size of the intended cpc is established by the selection of the wiring system to be used – i.e. steel-wire armoured cables, steel conduit, mineral insulated cables, etc. If the cross-sectional area of the armour, conduit, etc., is found to comply with the requirements of Table 54.7 then calculation of the thermal capacity is not usually necessary.

D2   Steel conduit and trunking

For some types of cpc, e.g. steel conduit, a check to prove compliance with Table 54.7 is quite simple and minimum cross-sectional areas of heavy gauge steel conduit and steel trunkings are listed in Table D5. Taking the value of k_1 as 115 for thermoplastic from Table 43.1 and k_2 as 47 from Table 54.5, it can be shown that heavy gauge steel conduit is a suitable protective conductor for all sizes of conductor that can be drawn into it in accordance with Appendix A.

A well installed and maintained steel conduit or cable trunking system can provide an adequate cpc for a final circuit (Regulation 543.2.2). Unfortunately, conduit and trunking impedance data for the calculation of earth fault loop impedances of circuits utilizing conduit or trunking as a protective conductor is not readily available and so the practice of installing a separate cpc has arisen. This is usually uneconomical and unnecessary.

However, when a conduit or trunking is to be utilized as a protective conductor, all joints must be tight and secure and the system protected from corrosion. The cross-sectional area of the protective conductor is to be taken as the minimum of any variations of size along the route. Consideration must also be made of the possible further reduction in cross-sectional area at conduit running couplers and cable trunking couplers, with only metal-to-metal contact at bolt holes even when copper links are fitted. Flanged couplers may be necessary to provide adequate continuity at conduit connections onto trunking systems.

▼ **Table D5** Cross-sectional areas of steel conduit and trunking

Heavy Gauge Steel Conduit to BS 4568-1 (BS EN 61386-21)

Nominal diameter (mm)	Minimum steel cross-sectional area (mm²)
16	64.4
20	82.6
25	105.4
32	137.3

Steel Surface Trunking to BS EN 50085-1 (sample sizes)

Nominal size (mm × mm)	Minimum steel cross-sectional area without lid (mm²)
50 × 50	135
75 × 75	243
100 × 50	216
100 × 100	324
150 × 100	378

Steel Underfloor Trunking to BS 4678-2 (sample sizes)

Nominal size (mm × mm)	Minimum steel cross-sectional area without lid (mm²)
75 × 25	118
100 × 50	142
100 × 100	213
150 × 100	284

The requirements of Regulation 543.3, as they apply, must be complied with in respect of the preservation of electrical continuity, for a protective conductor common to several circuits. The cross-sectional area must either be calculated to provide adequate capacity for the most onerous conditions of any of the several circuits, or sized from Table 54.7 to correspond with the largest line conductor of the several circuits. Therefore, a metal conduit or trunking could serve as a cpc to all the group of circuits it encloses.

Formulae for the calculation of the resistance and inductive reactance values of steel conduit and steel ducting and trunking are published in Guidance Note 6.

D3   Steel-wire armoured cables

If a BS EN 60269-2 or BS 88 fuse or a circuit-breaker is selected for overload protection of thermoplastic/SWA/thermoplastic cable to BS 6346 or XLPE/SWA/thermoplastic cable to BS 5467, then for an earth fault disconnection time of up to 5 s the steel armouring will have a sufficient cross-sectional area to comply with Regulation 543.1.3 with two exceptions. These exceptions are two-core 240 mm² and 300 mm² XLPE/SWA/thermoplastic (PVC) to BS 5467.

From I, t and k, the required minimum cross-sectional area of wire armour, S, can be calculated and compared with the area provided by the cable. The cross-sectional areas of steel-wire armour on thermosetting insulated cables to BS 5467 and thermoplastic insulated cables to BS 6346 are given in Tables 12E and 12F of the *Commentary on IEE Wiring Regulations*.

It will be found that, if the protective device is one of the standard types mentioned in the Regulations and if its rating is not greater than the current rating of the cable, the cross-sectional area of the armour is greater than required to comply with the Regulations. This assumes that the earth fault loop impedance of the circuit is such as to ensure that the device will operate in not more than 5 s on the occurrence of a fault of negligible impedance.

▼ **Table D6** Copper conductor, thermoplastic insulated steel-wire armoured cables to BS 6346 and fuses to BS 88

| Fuse rating | Required area of SWA (5 s disconnection) | Smallest cable providing required area of SWA | | | |
| | | 2-core | 3-core | 4-core (equal) | 4-core (red. neut.) |
A	mm²	mm²	mm²	mm²	mm²
*63	14.3	1.5 (20 A)	1.5 (18 A)	1.5 (18 A)	–
80	21	4 (37 A)	4 (31 A)	2.5 (24 A)	–
100	28	10 (69 A)	6 (42 A)	4 (31 A)	–
125	35	10 (69 A)	6 (42 A)	4 (31 A)	–
160	46	16 (90 A)	16 (77 A)	10 (58 A)	–
200	61	25 (121 A)	25 (102 A)	16 (77 A)	–
250	77	70 (220 A)	70 (220 A)	50 (155 A)	35 (125 A)
315	107	95 (270 A)	70 (190 A)	50 (155 A)	70 (190 A)
400	133	120 (310 A)	95 (230 A)	70 (190 A)	70 (190 A)
500	178	185 (410 A)	150 (310 A)	120 (310 A)	120 (270 A)
630	224	240 (485 A)	185 (350 A)	150 (310 A)	150 (310 A)
800	350	–	–	400 (550 A)	400 (550 A)

* For BS EN 60269-2 or BS 88 fuses of 6 to 50 A ratings the required armour area is, of course, less than that required for the 63 A fuse, and the cables with 1.5 mm² conductors (the smallest in BS 6346) provide areas exceeding the required minima.

▼ **Table D7** Copper conductor, thermoplastic insulated steel-wire armoured cables to BS 6346 and circuit-breakers to BS 3871 and BS EN 60898 (XLPE cables to BS 5467 also comply with the same requirements)

Circuit-breaker designation	Required area of SWA (5 s disconnection) mm²	Smallest cable providing required area of SWA		
		2-core mm²	3-core mm²	4-core mm²
*30/32 A Type 1	6.1	1.5 (20 A)	1.5 (18 A)	1.5 (18 A)
30/32 A Type 2 or Type B	10.7	1.5 (20 A)	1.5 (18 A)	1.5 (18 A)
30/32 A Type 3 or Type C	13.7	1.5 (20 A)	1.5 (18 A)	1.5 (18 A)
50 A Type 1	10.2	1.5 (20 A)	1.5 (18 A)	1.5 (18 A)
50 A Type 2 or Type B	17.8	1.5 (20 A)	1.5 (18 A)	1.5 (18 A)
50 A Type 3 or Type C	21.8	4 (37 A)	4 (31 A)	4 (31 A)

* For 5 to 20 A circuit-breakers of all types to BS 3871 and 6 to 25 A for BS EN 60898, the 1.5 mm² cables provide armour areas exceeding the required minima.

▼ **Table D8** Copper conductor, 90 °C thermosetting insulated steel-wire armoured cables to BS 5467 and fuses to BS 88

Fuse rating A	Required area of SWA (5 s disconnection) mm²	Smallest cable providing required area of SWA			
		2-core mm²	3-core mm²	4-core (equal) mm²	4-core (red. neut.) mm²
*125	28.2	16 (114 A)	16 (100 A)	16 (100 A)	–
160	37	16 (114 A)	16 (100 A)	16 (100 A)	–
200	50	35 (190 A)	25 (133 A)	25 (133 A)	25 (133 A)
250	62	50 (228 A)	25 (133 A)	25 (133 A)	25 (133 A)
315	87	95 (356 A)	70 (247 A)	50 (195 A)	50 (195 A)
400	108	95 (356 A)	95 (304 A)	70 (247 A)	70 (247 A)
500	145	185 (540 A)	150 (408 A)	95 (304 A)	95 (304 A)
630	182	185 (540 A)	150 (408 A)	120 (350 A)	150 (408 A)
800	286	–	–	240 (550 A)	300 (627 A)

* For fuses to BS EN 60269-2 or BS 88 of lower rating than 125 A, cables with 16 mm² conductors (the smallest size in BS 5467) provide armour areas exceeding the required minima.

Tables are also given below identifying which of the standard sizes of steel-wire armoured cables can be utilized to comply with Table 54.7.

▼ **Table D9** Thermoplastic /SWA/thermoplastic cables to BS 6346. Insulation operating at 70 °C

Minimum from Table 54.7, with $k_1 = 115$ (103 for 400 mm^2) from Table 43.1 and $k_2 = 46$ from table 54.4

Conductor csa (mm^2)	Minimum csa of steel cpc required to comply with Table 54.7 (mm^2)	Actual armour csa from BS 6346 (mm^2)		
		2-core	3-core	4-core
1.5	3.4	15	16	17
2.5	5.7	17	19	20
4	9.0	21	23	35
6	13.6	24	36	40
10	22.6	41	44	49
16	36.1	46	50	72
25	36.1	60	66	76
35	36.1	66	74	84
50	56.4	74	84	122
70	79.0	84	119	138
95	107.2	122	138	160
120	135.3	(131)	150	220
150	169.2	(144)	211	240
185	208.6	(201)	230	265
240	270.6	(225)	(260)	299
300	338.3	(250)	(289)	(333)
400	403.9	(279)	(319)	467

(-) Indicates that the cable does not comply with Table 54.7. Therefore, the cpc size must be confirmed by calculation as indicated by Regulation 543.1.3, or a supplementary cpc of the full conductor size must be installed.

▼ **Table D10a** 90 °C thermosetting insulated steel-wire armoured cables to BS 5467 and BS 6724. Insulation operating at 90 °C

Minimum from Table 54.7, with $k_1 = 143$ from Table 43.1 and $k_2 = 46$ from table 54.4

Conductor csa (mm²)	Minimum csa of steel cpc required to comply with Table 54.7 (mm²)	Actual armour csa from BS 5467 and BS 6724 (mm²)		
		2-core	3-core	4-core
1.5	4.7	15	16	17
2.5	7.8	17	19	20
4	12.5	19	21	23
6	18.7	22	23	36
10	31.1	(26)	39	43
16	49.8	(41)	(44)	(49)
25	49.8	(42)	62	70
35	49.8	62	70	80
50	77.8	(68)	78	90
70	108.8	(80)	(90)	131
95	147.7	(113)	(128)	(147)
120	186.6	(125)	(141)	206
150	233.2	(138)	(201)	(230)
185	287.6	(191)	(220)	(255)
240	373.1	(215)	(250)	(289)
300	466.3	(235)	(269)	(319)
400	521.7	(265)	(304)	(452)*

* Not applicable to BS 6724.

(-) Indicates that the cable does not comply with Table 54.7. Therefore, the cpc size must be confirmed by calculation as indicated by Regulation 543.1.3, or a supplementary cpc of the full conductor size must be installed.

There are times when an XLPE insulated cable, normally rated at 90 °C, may be run alongside thermoplastic insulated cables operating at 70 °C. Under such circumstances the 90 °C thermosetting insulated cable must be limited to the lower operating temperature of 70 °C and Table D10b should be used to determine if the armour complies with Table 54.7.

▼ **Table D10b** 90 °C thermosetting insulated steel-wire armoured cables to BS 5467 and BS 6724. Insulation operating at 70 °C

Minimum from Table 54.7, with $k_1 = 115$ from Table 43.1 and $k_2 = 51$ from table 54.4

Conductor csa (mm²)	Minimum csa of steel cpc required to comply with Table 54.7 (mm²)	Actual armour csa from BS 5467 and BS 6724 (mm²)		
		2-core	3-core	4-core
1.5	3.4	16	17	18
2.5	5.7	17	19	20
4	9.0	19	21	23
6	13.6	22	23	36
10	22.6	26	39	43
16	36.1	41	44	49
25	36.1	42	62	70
35	36.1	62	70	80
50	56.4	68	78	90
70	79.0	80	90	131
95	107.2	113	128	147
120	135.3	(125)	141	206
150	169.2	(138)	201	230
185	208.6	(191)	220	255
240	270.6	(215)	(250)	289
300	388.3	(235)	(269)	(319)
400	451.0	(265)	(304)	452*

* Not applicable to BS 6724.

(-) Indicates that the cable does not comply with Table 54.7. Therefore, the cpc size must be confirmed by calculation as indicated by Regulation 543.1.3, or a supplementary cpc of the full conductor size must be installed.

D4 Mineral insulated cable with copper sheath

Cross-sectional areas of the copper sheath of light duty and heavy duty mineral insulated cables to BS EN 60702-1 are given in Table D11 in which compliance with Table 54.7 is indicated.

D

▼ **Table D11** Cross-sectional areas of the copper sheath of light and heavy duty mineral insulated cables to BS EN 60702-1

Cable size reference	Effective sheath area (mm^2)
500 V Grade (light duty)	
2L1	5.4
2L1.5	6.3
2L2.5	8.2
2L4	10.7
3L1	6.7
3L1.5	7.8
3L2.5	9.5
4L1	7.7
4L1.5	9.1
4L2.5	11.3
7L1	10.2
7L1.5	11.8
7L2.5	15.4
750 V Grade (heavy duty)	
1H6.0	8.0
1H10	(9.0)
1H16	(12.0)
1H25	(15.0)
1H35	(18.0)
1H50	(22.0)
1H70	(27.0)
1H95	(32.0)
1H120	(37.0)
1H150	(44.0)
1H185	(54.0)
1H240	(70.0)
2H1.5	11
2H2.5	13
2H4	16
2H6	18
2H10	24
2H16	30
2H25	38
3H1.5	12
3H2.5	14
3H4	17

▼ **Table D11** continued

Cable size reference	Effective sheath area (mm^2)
3H6	20
3H10	27
3H16	34
3H25	42
4H1.5	14
4H2.5	16
4H4	20
4H6	24
4H10	30
4H16	39
4H25	49
7H1.5	18
7H2.5	22
12H2.5	34
19H2.5	37

(-) Indicates that the cable does not comply with Table 54.7. Therefore, the cpc size must be confirmed by calculation as indicated by Regulation 543.1.3, or a supplementary cpc of the full conductor size must be installed.

523.201   It can be seen that all multicore sizes comply with Table 54.7 but, with the exception of 1H6, the heavy duty single-core micc cables do not. Although these cables singularly do not comply, usually single-core cables would be run in pairs for a single-phase circuit, or three or four for a three-phase circuit. The sheath sections, if bonded together, would then sum together and comply for some conductors. (See Regulation 523.201 also.)

D5   Thermoplastic (PVC) insulated and sheathed cables to BS 6004

These common wiring cables have a cpc of a smaller size than the line conductor for all sizes excepting 1.0 mm^2. Consequently, only the 1.0 mm^2 cables (6241Y, 6242Y and 6243Y) will comply with Table 54.7. The standard conductor sizes of these cables are identified in Table E1 of Appendix E.

Consequent upon this, calculations are required for the cpc in circuits in which this type of cable is used, usually in domestic installations. *The On-Site Guide* Table 7.1(i) contains calculated maximum circuit lengths for various arrangements of final circuits and types of protective device.

Table 52.1   Similar cable size constructions apply to certain thermosetting cables to BS 7211 BS EN 50525. These cables, however, can have a conductor operating temperature of up to 90 °C. Table 52.1 of BS 7671 requires that equipment and accessories connected to such conductors operating at temperatures exceeding 70 °C must be suitable for the resulting higher temperature at the connection.

Domestic accessories and equipment are not usually suitable for operation at such higher temperatures and consequently cables having a rated conductor operating temperature exceeding 70 °C cannot be used at their full rating and should be considered as thermoplastic insulated and sheathed types.

Appendix E

Resistance and impedance of copper and aluminium conductors under fault conditions

To check compliance with Regulation 434.5.2 and/or Regulation 543.1.3, i.e. to evaluate the equation $S^2 = I^2 t/k^2$, it is necessary to establish the impedances of the circuit conductors to determine the fault current I and hence the protective device disconnection time t.

Similarly, in order to design circuits for compliance with the limiting values of earth fault loop impedance given in Tables 41.2 to 41.4 of BS 7671, it is necessary to establish the relevant impedances of the circuit conductors concerned. (See Guidance Note 5 for more information on this subject.)

Where the circuit overcurrent protective device characteristics comply with those given in Appendix 3 of BS 7671, the increase in circuit conductor temperature at fault has been deemed to be taken into account and only the normal conductor operating temperature needs to be considered.

The equations given in Regulations 434.5.2 and 543.1.3 assume a constant value of fault current but in practice that current changes during the period of the fault because, due to the rise in temperature, the conductor resistance increases.

The rigorous method for taking into account the changing character of the fault current is too complicated for practical use and over the range of temperatures encountered in BS 7671 a sufficiently accurate method is to calculate conductor impedances based on the full load current temperature of the conductor (i.e. 70 °C for thermoplastic insulated conductors and 90 °C for thermosetting insulated conductors) for devices given in Appendix 3 of BS 7671.

Appendix 3 of BS 7671 does not include all types of device, for example circuit-breakers (MCCBs) to BS EN 60947-2:2006. Earth fault loop impedances are not tabulated in BS 7671 for MCCBs nor are their characteristics shown in Appendix 3. It may be that these devices do comply with the current regulation requirements but there is a wide variety of devices with adjustable ranges and characteristics and if there is any doubt, the temperature rise of the average of the initial and final fault temperatures should be utilized. The device manufacturer's advice may be required.

Most manufacturers will, however, provide details of the maximum allowable impedance values for both 0.4 s and 5 s disconnection.

The following table of conductor resistances, Table E1, is limited to conductor cross-sectional areas up to and including 35 mm², i.e. to conductors having negligible inductive reactance. For larger cables the reactance is not negligible. This reactance is independent of temperature and depends on conductor size and cable make-up and it may be necessary to obtain information from the manufacturer as regards the resistive and reactive components of the impedance of the cables it is intended to use.

Table E1 gives values of resistance of cables up to 35 mm^2. Values for single conductors are given as well as loop ($R_1 + R_2$) resistances for various combinations, particularly the reduced size cpcs used in flat twin and earth cable.

Table E2 gives the multipliers to be applied to the values given in Table E1 for the purpose of calculating the resistance, at maximum permissible operating temperature, of line conductors and/or circuit protective conductors, in order to determine compliance with, as applicable:

(a) earth fault loop impedances of Tables 41.2 to 41.4
(b) the equation in Regulation 434.5.2
(c) the equation in Regulation 543.1.3
(d) earth fault loop impedances of Appendix D in this or other Guidance Notes.

Where it is known that the actual operating temperature under normal load is less than the maximum permissible value for the type of cable insulation concerned (as given in the tables of current-carrying capacity), the multipliers given in Table E2 may be reduced, see Appendix 4 of BS 7671 for further details.

▼ **Table E1** Values of resistance/metre for copper and aluminium conductors and of
$(R_1 + R_2)$/metre in milliohms/metre

Cross-sectional area (mm²)		Resistance/metre or $(R_1 + R_2)$/metre (mΩ/m) at 20 °C		Resistance/metre or $(R_1 + R_2)$/metre (mΩ/m) at 70 °C	
Line conductor	Protective conductor	Plain copper	Aluminium	Plain copper	Aluminium
1	–	18.10		21.72	
1*	1	36.20		43.44	
1.5	–	12.10		14.52	
1.5*	1	30.20		36.24	
1.5	1.5	24.20		29.04	
2.5	–	7.41		8.89	
2.5	1	25.51		30.61	
2.5*	1.5	19.51		23.41	
2.5	2.5	14.82		17.78	
4	–	4.61		5.53	
4*	1.5	16.71		20.05	
4	2.5	12.02		14.42	
4	4	9.22		11.06	
6	–	3.08		3.70	
6*	2.5	10.49		12.59	
6	4	7.69		9.23	
6	6	6.16		7.39	
10	–	1.83		2.20	
10*	4	6.44		7.73	
10	6	4.91		5.89	
10	10	3.66		4.39	
16	–	1.15	1.91	1.38	2.29
16*	6	4.23	–	5.08	–
16	10	2.98	–	3.58	–
16	16	2.30	3.82	2.76	4.58
25	–	0.727	1.20	0.87	1.44
25	10	2.557	–	3.07	–
25	16	1.877	–	2.25	–
25	25	1.454	2.40	1.74	2.88
35	–	0.524	0.868	0.63	1.04
35	16	1.674	2.778	2.01	3.33
35	25	1.251	2.068	1.50	2.48
35	35	1.048	1.736	1.26	2.08

* Identifies copper line/protective conductor combination that complies with Table 5 or Table 6 of BS 6004:2000 for thermoplastic insulated and sheathed single, twin or three-core and cpc cables (i.e. 6241Y, 6242Y or 6243Y cables) and similar cable constructions for thermosetting cables to BS 7211:1998.

▼ **Table E2** Operating temperature multipliers to Table E1

Insulation material	Multiplier	
	54.2	54.3
70 °C thermoplastic	1.04	1.20
90 °C thermoplastic	1.04	1.28
90 °C thermosetting	1.04	1.28

The multipliers given in Table E2 are based on the simplified formula given in BS EN 60228 for both copper and aluminium conductors, namely, that the resistance temperature coefficient, k_t, is 0.004 per °C at 20 °C.

Table 54.2 **54.2** applies where the protective conductor is not incorporated or bunched with cables, or for bare protective conductors in contact with cable covering (assumed initial temperature 30 °C).

Table 54.3 **54.3** applies where the protective conductor is a core in a cable or is bunched with cables (assumed initial temperature 70 °C or greater).

E1 Verification

For verification of earth fault loop impedances at installation completion the person carrying out testing will need the maximum values of the line and protective conductor resistances of a circuit at the ambient temperature expected during tests. This may be different from the reference temperature of 20 °C used for Table E1.

Table E3 gives correction factors that may be applied to the 20 °C Table E1 values to take account of the ambient temperature (for test purposes only). See also Guidance Note 3: *Inspection & Testing*.

Table E3 Ambient temperature multipliers to Table E1

Expected ambient temperature (°C)	Correction factor
5	0.94
10	0.96
15	0.98
25	1.02

Tables have been published in the *On-Site Guide* to provide maximum circuit earth fault loop impedances at 20 °C ready for test comparison.

E2 Mineral insulated cable with copper sheath

Values of conductor resistance (R_1) and sheath resistance (R_2) per metre for copper sheathed light duty and heavy duty mineral insulated cables to BS EN 60702 are given in Table E4.

▼ **Table E4** Values of R_1 and R_2 for mineral insulated cable with copper sheath (in Ω/km)

Cable ref.	R_1 Conductor resistance 20 °C	R_2 Sheath resistance 20 °C	R_1 Conductor resistance	R_2 Sheath resistance	R_1 Conductor resistance	R_2 Sheath resistance
			Exposed to touch 70 °C sheath		Not exposed to touch 105 °C sheath	
500 V Grade (light duty)						
2L1	18.1	3.95	21.87	4.47	24.5	4.84
2L1.5	12.1	3.35	14.62	3.79	16.38	4.1
2L2.5	7.41	2.53	8.95	2.87	10.03	3.1
2L4	4.61	1.96	5.57	2.22	6.24	2.4
3L1	18.1	3.15	21.87	3.57	24.5	3.86
3L1.5	12.1	2.67	14.62	3.02	16.38	3.27
3L2.5	7.41	2.23	8.95	2.53	10.03	2.73
4L1	18.1	2.71	21.87	3.07	24.5	3.32
4L1.5	12.1	2.33	14.62	2.64	16.38	2.85
4L2.5	7.41	1.85	8.95	2.1	10.03	2.27
7L1	18.1	2.06	21.87	2.33	24.5	2.52
7L1.5	12.1	1.78	14.62	2.02	16.38	2.18
7L2.5	7.41	1.36	8.95	1.54	10.03	1.67
750 V Grade (heavy duty)						
1H10	1.83	2.23	2.21	2.53	2.48	2.73
1H16	1.16	1.81	1.4	2.05	1.57	2.22
1H25	0.727	1.4	0.878	1.59	0.984	1.72
1H35	0.524	1.17	0.633	1.33	0.709	1.43
1H50	0.387	0.959	0.468	1.09	0.524	1.18
1H70	0.268	0.767	0.324	0.869	0.363	0.94
1H95	0.193	0.646	0.233	0.732	0.261	0.792
1H120	0.153	0.556	0.185	0.63	0.207	0.681
1H150	0.124	0.479	0.15	0.542	0.168	0.587
1H185	0.101	0.412	0.122	0.467	0.137	0.505
1H240	0.0775	0.341	0.0936	0.386	0.105	0.418
2H1.5	12.1	1.9	14.62	2.15	16.38	2.33
2H2.5	7.41	1.63	8.95	1.85	10.03	2
2H4	4.61	1.35	5.57	1.53	6.24	1.65
2H6	3.08	1.13	3.72	1.28	4.17	1.38
2H10	1.83	0.887	2.21	1.005	2.48	1.09
2H16	1.16	0.695	1.4	0.787	1.57	0.852
2H25	0.727	0.546	0.878	0.618	0.984	0.669
3H1.5	12.1	1.75	14.62	1.98	16.38	2.14

E

Appendix F

Selection and erection of wiring systems

F1   General

511.1   To conform to the requirements of BS 7671, Regulation 511.1 requires that all electrical equipment, including wiring systems, must utilize cables complying with the relevant requirements of the applicable British or Harmonized Standard.

Alternatively, if equipment complying with a foreign national standard based on an IEC Standard is to be used, the designer or other person responsible for specifying the installation must verify that any differences between that standard and the corresponding British or Harmonized Standard will not result in a lesser degree of safety than that afforded by compliance with the British Standard.

Where equipment to be used is not covered by a British or Harmonized Standard or is used outside the scope of its standard, the designer or other person responsible for specifying the installation must satisfy themselves and confirm that the equipment provides the same degree of safety as that afforded by compliance with the Regulations.

Part 2   A 'wiring system' is defined in Part 2 of BS 7671 as 'An assembly made up of cable or busbars and parts which secure and, if necessary, enclose the cable or busbars'. This can be read to only mean factory-made systems but it is intended to cover all cable types.

Cables are also identified with a voltage grade, to identify the maximum system working voltage for which they are suitable. Conduit wiring cables (6491X), etc., are designated 450/750 V and are harmonized within CENELEC under HD 21 and HD 22. Wiring cables such as flat twin and earth (6242Y) are designated 300/500 V and are not harmonized but are constructed to a British Standard. Armoured cables are designated 600/1000 V and are not harmonized but are also constructed to a British Standard. There is no difference in utilizing types of any of these designations on the UK 230/400 V supply system.

BS 7540:2005 (series) *Electric cables. Guide to use for cables with a rated voltage not exceeding 450/750 V* gives installation application advice.

▼ **Table F1** Comparison of harmonized cable types to BS 6004

Type	British Standard table no.	Code designation	Conductor class	Voltage grade (V)	Number of cores	csa range (mm^2)	Temperature (°C)	
							Installation (minimum)	Storage (maximum)
Thermoplastic insulated non-sheathed	1	H07V-U	1	450/750	1	1.5–16	+5	+40
		H07V-R	2			1.5–630		
		H07V-K	5			1.5–240		
Thermoplastic insulated non-sheathed	2	H05V-U	1	300/500	1	0.5–1.0	+5	+40
Thermoplastic insulated and sheathed	3		1/2	300/500	2–5	1.5–35	+5	+40
Thermoplastic insulated and sheathed	4	6181Y 6192Y 6193Y	1/2	300/500	1–3	1.0–35	+5	+40
Thermoplastic insulated and sheathed	5	6242Y 6243Y	1/2	300/500	2-3	1.0–16	+5	+40
Thermoplastic insulated and sheathed	6	6192Y 6242Y	2	300/500	2	1.5–2.5	+5	+40

▼ **Table F2** Designation system for cables complying with the European Harmonization Standard

Cable Reference

[1 2] [3 4 5] - [6] [7 8 9]

1 Basic Standards

H   Harmonized Standards
A   Authorized National Standards (derived from a harmonized cable standard)
N   Non-Authorized National Standards

2 Rated voltage

03   300/300 V
05   300/500 V
07   450/750 V
1   600/1000 V

3 & 4 Insulation and sheathing material

B   Ethylene propylene rubber (EPR)
E   Polyethylene (PE), low density (LDPE)
E2   Polyethylene, high density (HDPE)
E4   Polytetrafluoroethylene (PTFE)
E6   Ethylene tetrafluoroethylene (ETFE)
E7   Polypropylene (PP)
G   Ethylene vinyl acetate (EVA)
J   Glass fibre braid (GFB)
N   Polychloroprene (PCP)
N4   Chlorosulphonated polyethylene (CSP)
Q   Polyurethane (PU)

Q2	Polyethylene terephthalate (PETP)
Q4	Polyamide (PA)
R	Natural rubber
S	Silicone
T	Textile braid
V	Polyvinyl chloride (PVC)
V2	Heat-resistant polyvinyl chloride (HR PVC)
X	Cross-linked polyethylene (XLPE)

5 Special construction and shapes

H	Flat construction with divisible cores
H2	Flat construction, non-divisible core
H5	Two or more cores twisted together, non-sheathed

6 Type of conductor

A	Aluminium
	Copper (no code letter)
F	Flexible for movable installations (Class 5 IEC 228)
H	Highly flexible for movable installations (Class 6 IEC 228)
K	Flexible for fixed installations (Class 5 IEC 228)
R	Stranded (Class 2 IEC 228)
U	Solid (Class 1 IEC 228)
Y	Tinsel

7 Number of cores

8 Protective conductor

X	Without protective core
G	With protective core

9 Nominal cross-sectional area of conductors in mm²

Additional designations

Concentric conductors and screens

A	Concentric aluminium conductor
C	Concentric copper conductor
A7	Aluminium/Laminate screen
C4	Overall copper braid screen
C5	Cores individually copper braid screen
C7	Lapped copper (wire, tape or strip) screen

Special components

D3	Central strainer (textile or metallic)
D5	Central filler (not load bearing)

Armours

Z2	Steel wire armour
Z3	Flat steel wire armour
Z4	Steel tape armour
Z5	Steel wire braid

▼ **Table F3** Commonly used cable type reference numbers

Reference no.	Cable type
0361	Welding cable, tinned copper conductor, vulcanised rubber insulation/tough rubber sheath. BS 638
2093Y	3-core, thermoplastic heat resisting insulation/thermoplastic heat resisting sheathed. Circular, 300/300 V BS 6500, flexible cord
218*Y	Circular, thermoplastic insulation/thermoplastic sheathed. 300/300 V BS 6500, flexible cord
2192Y	2-core, thermoplastic insulation/thermoplastic sheathed flat 300/300 V. BS 6500, flexible cord
2213	3-core, vulcanised rubber insulation/tough rubber sheathed/ cotton braid, 300/300 V BS 6500, flexible cord
318*	Circular, vulcanised rubber insulation/tough cord rubber sheathed, 300/500 V BS 6500, flexible cord
318*Y	Circular, thermoplastic insulation/thermoplastic sheathed 300/500 V BS 6500, flexible cord
318*P	Circular, vulcanised rubber insulation/PCP sheathed, 300/500 V BS 6500, flexible cord
318*TQ	Circular, EPR insulation/CSP sheathed 300/500V BS 6500, flexible cord
618*Y	Circular, thermoplastic insulation/thermoplastic sheathed 1–35 mm^2 300/500 V BS 6004 50–630 mm^2 gen in acc BS 6346, wiring cable
619*Y	Flat, thermoplastic insulation/thermoplastic sheathed 300/500 V BS 6004, wiring cable
624*Y	Flat, thermoplastic insulation – laid up in parallel with a bare CPC within one interstice/thermoplastic sheathed 300/500 V BS 6004, wiring cable
6491X	Single-core, thermoplastic insulation 1–630 mm^2. 450/750 V BS 6004
657*TQ	Circular, EPR insulation/CSP sheathed 600/1000 V BS 6883, ship-wiring cable. (circular conductors)
657*TQ(S)	As 657*TQ but with shaped conductors
658*TQ	Circular, EPR insulation/CSP bedding/galv. mild steel wire braid armour/CSP sheathed 600/1000 V BS 6883, ship wiring cable (circular conductors)
658*TQ(S)	As 658*TQ but with shaped conductors
694*X	Circular, thermoplastic insulation/thermoplastic bedded/single steel wire armoured/thermoplastic sheathed 600/1000 V BS 6346, mains wiring cable
694*X(S)	As 694*X but with shaped conductors
H07V-R	Single-core, thermoplastic insulation, 1.5–630 mm^2, (rigid strand) 450/750 V to BS 6004
H07V-U	Single-core, thermoplastic insulation, 1.5–2.5 mm^2, (solid) 450/750 V to BS 6004

▼ **Table F3**   *continued*

Reference no.	Cable type
H05V-U	Single-core, thermoplastic insulation, 1.0 mm² (solid) 300/500 V to BS 6004
H05RR-F	Circular, vulcanised rubber insulation/tough rubber sheathed. 300/500 V to BS 6500
H03VV-F	Circular, thermoplastic insulated/thermoplastic sheathed, 300/300 V to BS 6500
H03VVH2-F	Flat 2-core, thermoplastic insulated/thermoplastic sheathed, 300/300 V to BS 6500
H05VV-F	Circular, thermoplastic insulated/thermoplastic sheathed, 300/500 V to BS 6500
H05VVH2-F	Flat, 2-core, thermoplastic insulated/thermoplastic sheathed, 300/500 V to BS 6500
H05V-K	Single-core, thermoplastic insulation 0.5–1.0 mm² 300/500 V to BS 6500
H07V-K	Single-core, thermoplastic insulation, 1.5–2.5 mm², 450/750 V to BS 6004
H07RN-F	Circular, rubber insulated/rubber sheathed, 450/750 V to BS 6007

Notes:

* = number of cores (e.g. 2183Y is 3-core)

(AL) = aluminium conductors. To be added after other suffix, if used, e.g. 6491X(AL) and 6242Y(AL)

(S) = shaped conductors

(Reproduced with kind permission of Anixter (UK) Ltd.)

▼ **Table F4** Applications of cables for fixed wiring

Type of cable	Uses	Comments
Thermoplastic, thermosetting or rubber insulated non-sheathed	In conduit, cable ducting or trunking	(i) intermediate support may be required on long vertical runs (ii) 70 °C maximum conductor temperature for normal wiring grades — including thermosetting types[4] (iii) cables run in PVC conduit shall not operate with a conductor temperature greater than 70 °C[4]
Flat thermoplastic or thermosetting insulated and sheathed	(i) general indoor use in dry or damp locations. May be embedded in plaster (ii) on exterior surface walls, boundary walls and the like (iii) overhead wiring between buildings[6] (iv) underground in conduits or pipes (v) in building voids or ducts formed in-situ	(i) additional protection may be necessary where exposed to mechanical stresses (ii) protection from direct sunlight may be necessary. Black sheath colour is better for cables in sunlight (iii) see note d (iv) unsuitable for embedding directly in concrete (v) may need to be hard drawn (HD) copper conductors for overhead wiring[6]
Split-concentric thermoplastic insulated and sheathed	General	(i) additional protection may be necessary where exposed to mechanical stresses (ii) protection from direct sunlight may be necessary. Black sheath colour is better for cables in sunlight
Mineral insulated	General	With overall PVC covering where exposed to the weather or risk of corrosion, or where installed underground, or in concrete ducts
Thermoplastic or XLPE insulated, armoured, thermoplastic sheathed	General	(i) additional protection may be necessary where exposed to mechanical stresses (ii) protection from direct sunlight may be necessary. Black sheath colour is better for cables in sunlight
Paper-insulated, lead sheathed and served	General, for main distribution cables	With armouring where exposed to severe mechanical stresses or where installed underground

Notes:

522.8.10 **(a)** The use of cable covers (preferably conforming to BS 2484) or equivalent mechanical protection is desirable for all underground cables which might otherwise subsequently be disturbed. Route marker tape should also be installed, buried just below ground level.

(b) Cables having thermoplastic insulation or sheath should preferably be installed only when the ambient temperature is above 0 °C and has been for the preceding 24 hours. Where they are to be installed during a period of low temperature, precautions should be taken to avoid risk of mechanical damage during handling. A minimum ambient temperature of 5 °C is advised in BS 7540-2:2005 for some types of thermoplastic insulated and sheathed cables.

(c) Cables should be suitable for the maximum ambient temperature and should be protected from any excess heat produced by other equipment, including other cables.

(d) Thermosetting cable types (to BS 7211 or BS 5467) can operate with a conductor temperature of 90 °C. This must be limited to 70 °C when drawn into a conduit, etc., with thermoplastic insulated conductors or connected to electrical equipment (512.1.5 and Table 52.1), or when such cables are installed in plastic conduit or trunking.

(e) For cables to BS 6004, BS 6007, BS 7211, BS 6346, BS 5467 and BS 6724, further guidance may be obtained from those standards. Additional advice is given in BS 7540-2:2005 *Guide to use for cables with a rated voltage not exceeding 450/750 V* for cables to BS 6004, BS 6007 and BS 7211.

(f) Cables for overhead wiring between buildings must be able to support their self-weight and any imposed wind or ice/snow loading. A catenary support is usual but hard drawn copper types may be used.

▼ **Table F5** Applications of flexible cables to BS 6500:2000 and BS 7919:2001

Type of flexible cable	Uses
Light thermoplastic (PVC) insulated and sheathed flexible cord	Indoors in household or commercial premises in dry situations, for light duty.
Ordinary thermoplastic (PVC) insulated and sheathed flexible cord	Indoors in household or commercial premises, including damp situations, for medium duty. For cooking and heating appliances where not in contact with hot parts. For outdoor use other than in agricultural or industrial applications. For electrically powered hand tools.
60 °C thermosetting (rubber) insulated braided twin and three-core flexible cord	Indoors in household or commercial premises where subject only to low mechanical stresses.
60 °C thermosetting (rubber) insulated and sheathed flexible cord	Indoors in household or commercial premises where subject only to low mechanical stresses. For occasional use outdoors. For electrically powered hand tools.
60 °C thermosetting (rubber) insulated oil-resisting with flame-retardant sheath	For general use, unless subject to severe mechanical stresses. For use in fixed installations where protected by conduit or other enclosures.
90 °C thermosetting (rubber) insulated HOFR sheathed	General, including hot situations, e.g. night storage heaters, immersion heaters and boilers.
90 °C heat resisting thermoplastic (PVC) insulated and sheathed	General, including hot situations, e.g. for pendant luminaires.
150 °C thermosetting (rubber) insulated and braided	For use at high ambient temperatures. For use in or on luminaires.
185 °C glass-fibre insulated single-core twisted twin and three-core	For internal wiring of luminaires only and then only where permitted by BS 4533.
185 °C glass-fibre insulated braided circular	For dry situations at high ambient temperatures and not subject to abrasion or undue flexing. For the wiring of luminaires.

Notes:

(a) Cables having thermoplastic insulation or sheath should preferably not be used where the ambient temperature is consistently below 0 °C. Where they are to be installed during a period of low temperature, precautions should be taken to avoid risk of mechanical damage during handling.

(b) Cables should be suitable for the maximum ambient temperature and should be protected from any excess heat produced by other equipment, including other cables.

(c) For flexible cables to BS 6007, BS 6141 and BS 6500 further guidance may be obtained from those standards, or from BS 7540-2:2005 *Guide to use for cables with a rated voltage not exceeding 450/750 V.*

(d) When used as connections to equipment, flexible cables should be of the minimum practical length to minimize danger and, in any case, of such a length that allows the protective device to operate correctly.

(e) When attached to equipment, flexible cables should be protected against tension, crushing, abrasion, torsion and kinking, particularly at the inlet point to the electrical equipment. At such inlet points it may be necessary to use a device which ensures that the cable is not bent to an internal radius below that given in the appropriate part of Table 4 of BS 6007. Strain relief, clamping devices or cable guards should not damage the cable.

(f) Flexible cables should not be used under carpets or other floor coverings, or where furniture or other equipment may rest on them. Flexible cables should not be placed where there is a risk of damage from traffic passing over them.

(g) Flexible cables should not be used in contact with or close to heated surfaces especially if the surface approaches the upper thermal limit of the cable or cable.

F2   British Standards

511.1 Most cables in general use today are manufactured to a specific British Standard that covers the design, materials, manufacture, testing and performance of the cable. The designer of the installation has to select a cable type that is suitable for the performance required, and the relevant British Standard is usually specified.

The cable manufacturer will provide full details of cables and the standards to which they are manufactured. BS 7671 Appendix 4 indicates the relevant British Standards for each cable type in the current rating tables. A full listing of British Standards referred to in BS 7671 is provided in Appendix 1 of that Standard.

F3   Fire stopping

527.2 All cable routes that pass between building fire zones or areas must be adequately sealed against the transmission of flames and/or smoke between zones or areas.

The specification for such fire stopping is outside the scope of this Guidance Note but should provide fire resistance to a standard at least equal to the original element of the building construction. The specification for fire stopping should be detailed by the Project Architect or Designer for all services (i.e. cables, pipes, etc.) and should comply with the requirements of the Building Regulations, as applicable. Intumescent materials are commonly used.

527.2.3 All voids within ducts, trunking, busbar trunking systems, etc., should be filled, as well as the space around such ducts, trunkings and busbar trunking systems where they pass through walls, etc. It is not usual, however, to seal the inside of conduits, except in classified hazardous areas, as required by the relevant codes or standards.

Appendix G

Notes on methods of support for cables, conductors and wiring systems

This appendix describes methods of support for cables, conductors and wiring systems which satisfy the relevant requirements of Chapter 52 of BS 7671. The use of other methods is not precluded where specified by a competent person, having due regard for protection of the cable from mechanical damage. The methods described in this appendix make no specific provisions for fire or thermal protection of the cables. All cable installations and routings must be carefully considered and protection provided as required. The advice of the local authority and the Fire Officer must be taken as necessary on the project. The Project Architect and Designer must also comply with the CDM Regulations (see Guidance Note 4 also).

G1   Cables generally

Items **A** to **I** below are generally applicable to supports on structures which are subject only to vibration of low severity and a low risk of mechanical impact:

A   Non-sheathed cables, installed in conduit without further fixing of the cables, precautions taken against undue compression or other mechanical stressing of the insulation at the top of any vertical runs exceeding 5 m in length.

B   Cables of any type, installed in factory-made ducting or trunking without further fixing of the cables, vertical runs not exceeding 5 m in length without intermediate support.

C   Sheathed and/or armoured cables drawn into ducts formed in-situ in the building structure. The internal surfaces of the duct must be protected to prevent abrasion of the cables, especially when drawing-in. Ducts should also be adequately sealed against the spread of fire and smoke. Vertical runs not exceeding 5 m in length without intermediate support.

D   Sheathed and/or armoured cables installed in accessible positions, support by clips at spacings not exceeding the appropriate value stated in Table G1.

E   Cables of any type, resting without fixing in horizontal runs of ducts, conduits, factory-made cable ducting or trunking.

F   Sheathed and/or armoured cables in horizontal runs which are inaccessible and unlikely to be disturbed, resting without fixing on part of a building, the surface of that part being reasonably smooth.

G   Sheathed and/or armoured cables in vertical runs which are inaccessible and unlikely to be disturbed, supported at the top of the run by a clip and a rounded support of a radius not less than the appropriate value stated in Table G2.

H Sheathed cables without armour in vertical runs which are inaccessible and unlikely to be disturbed, supported by the method described in **G** above; the length of run without intermediate support not exceeding 2 m for a lead sheathed cable or 5 m for a thermosetting or thermoplastic sheathed cable.

I Thermosetting or thermoplastic sheathed cables, installation in conduit without further fixing of the cables, any vertical runs being in conduit of suitable size and not exceeding 5 m in length.

G2 Particular applications

559.5.1.201
559.5.2

J Flexible cables used as pendants, attachment to a ceiling rose or similar accessory and lampholder by the cable grip or other method of strain relief provided in the accessories (see Table 4F3A of BS 7671 for maximum mass supportable by flexible cables).

K Temporary installations and installations on construction sites, protection and supports so arranged that there is no appreciable mechanical strain on any cable termination or joint and the cables are protected from mechanical damage.

721.522.8.1.3

L Caravans, for sheathed flexible cables in inaccessible spaces such as ceiling, wall and floor spaces, support at intervals not exceeding 0.25 m for horizontal runs and 0.4 m for vertical runs. For horizontal runs of sheathed flexible cables passing through floor or ceiling joists in inaccessible floor or ceiling spaces, securely bedded in thermal insulating material, no further fixing is required. (See Regulation 721.521 for details of caravan wiring systems.)

G3 Overhead wiring

M Cables sheathed with thermosetting or thermoplastic, supported by a separate catenary wire, either continuously bound up with the cable or attached thereto at intervals not exceeding those stated in column 2 of Table G1 and the minimum height above ground being in accordance with Table G3.

N Support by a catenary wire incorporated in the cable during manufacture, the spacings between supports not exceeding those stated by the manufacturer and the minimum height above ground being in accordance with Table G3.

O Spans without intermediate support (e.g. between buildings) of multicore sheathed cable, with or without armour, terminal supports so arranged that no undue strain is placed upon the conductors or insulation of the cable, adequate precautions being taken against any risk of chafing of the cable sheath and the minimum height above ground and the length of such spans being in accordance with the appropriate values indicated in Table G3. Hard drawn (HD) copper conductors may be necessary for longer spans. Cable manufacturers' advice should be complied with.

P Bare or insulated conductors of an overhead line for distribution between a building and a remote point of utilization (e.g. another building) supported on insulators, the lengths of span and heights above ground having the appropriate values indicated in Table G3 or otherwise installed in accordance with the Electricity Safety, Quality and Continuity Regulations 2002 (as amended).

Q Spans without intermediate support (e.g. between buildings) and which are in situations inaccessible to vehicular traffic, insulated or multicore sheathed cables installed in heavy gauge galvanized steel conduit, the length of span and height above ground being in accordance with Table G3, provided that the conduit shall be earthed in accordance with Part 4 and 5 of BS 7671, must be securely fixed at the ends of the span and not be jointed in its span.

▼ **Table G1**   Spacings of supports for cables in accessible positions where the entire support is derived from the clips

Overall diameter of cable, d* (mm)	Maximum spacings of clips (mm)							
	Non-armoured thermosetting, plastics or lead sheathed cables				Armoured cables		Mineral insulated copper sheathed cables	
	Generally		In caravans					
	Horizontal†	Vertical†	Horizontal†	Vertical†	Horizontal†	Vertical†	Horizontal†	Vertical†
1	2	3	4	5	6	7	8	9
d ≤ 9	250	400	250	400	–	–	600	800
9 < d ≤ 15	300	400	as above		350	450	900	1200
15 < d ≤ 20	350	450	as above		400	550	1500	2000
20 < d ≤ 40	400	550	as above		450	600	2000	3000

Notes:

▶ For the spacing of supports for cables having an overall diameter exceeding 40 mm, or conductors of cross-sectional area 300 mm² and larger, the manufacturer's recommendations should be observed.

▶ The spacings given in this table are maxima and for good workmanship in certain circumstances they may need to be reduced.

* For flat cables taken as the dimension of the major axis.

† The spacings stated for horizontal runs may be applied also to runs at an angle of more than 30 ° from the vertical. For runs at an angle of 30 ° or less from the vertical, the vertical spacings are applicable.

▼ **Table G2** Minimum internal radii of bends in cables for fixed wiring

Insulation	Finish	Overall diameter*	Factor to be applied to overall diameter of cable to determine minimum internal radius of bend
Thermosetting or thermoplastic (circular, or circular stranded copper or aluminium conductors)	Non-armoured	Not exceeding 10 mm	3(2)†
		Exceeding 10 mm but not exceeding 25 mm	4(3)†
		Exceeding 25 mm	6
	Armoured	Any	6
Thermosetting or thermoplastic (solid aluminium or shaped copper conductors)	Armoured or non-armoured	Any	8
Mineral	Copper sheath with or without covering	Any	6‡
Flexible cables	Sheathed	Any	No specific provision but no tighter than the equivalent sized non-armoured cable**

Notes:

* For flat cables the diameter refers to the major axis.

† The figure in brackets relates to single-core circular conductors of stranded construction installed in conduit, ducting or trunking.

‡ For mineral insulated cables, the bending radius should normally be limited to a minimum of 6 times the diameter of the bare copper sheath, as this will allow further straightening and reworking if necessary. However, cables may be bent to a radius not less than 3 times the cable diameter over the copper sheath, provided that the bend is not reworked.

** Flexible cables can be damaged by too tight, or repeated bending.

▼ **Table G3** Maximum lengths of span and minimum heights above ground for overhead wiring between buildings, etc.

Type of system	Maximum length of span (m)	Minimum height of span above ground (m)†		
		In		
		At road crossings	In positions accessible to vehicular traffic, other than crossings	In positions inaccessible to vehicular traffic*
1	2	3	4	5
Cables sheathed with thermoplastic or having an oil-resisting and flame-retardant (HOFR) sheath without intermediate support.	3	5.8	5.8	3.5
Cables sheathed with thermoplastic or having an oil-resisting and flame-retardant (HOFR) sheath, in heavy gauge steel conduit of a diameter not less than 20 mm and not jointed in its span.	3	5.8	5.8	3
Bare or covered overhead lines supported by insulators without intermediate support.	30	5.8	5.8	5.2
Cables sheathed with thermoplastic or having an oil-resisting and flame-retardant (HOFR) sheath, supported by a catenary wire.	No limit	5.8	5.8	3.5
Overhead cable incorporating a catenary wire.	Subject to Item 14	5.8	5.8	3.5

Notes:

*   This column is not applicable in agricultural premises.

†   In some special cases, such as in yacht marinas or where large cranes are present, it will be necessary to increase the height of span above ground over the minimum given in the table. It is preferable to use underground cables in such locations.

▶   Schedule 2 of the Electricity Safety, Quality and Continuity Regulations 2002 should be consulted for the minimum height above ground of overhead lines.

G4 Conduit and cable trunking

R Rigid conduit supported in accordance with Table G4.

S Cable trunking supported in accordance with Table G5.

T Conduit embedded in the material of the building, suitably treated against corrosion if necessary.

U Pliable conduit embedded in the material of the building or in the ground, or supported in accordance with Table G4.

▼ **Table G4** Spacings of supports for conduits

| Nominal diameter of conduit, d (mm) | Maximum distance between supports (m) | | | | | |
| | Rigid metal | | Rigid insulating | | Pliable | |
	Horizontal	Vertical	Horizontal	Vertical	Horizontal	Vertical
d ≤ 16	0.75	1.0	0.75	1.0	0.3	0.5
16 < d ≤ 25	1.75	2.0	1.5	1.75	0.4	0.6
25 < d ≤ 40	2.0	2.25	1.75	2.0	0.6	0.8
0.8 d > 40	2.25	2.5	2.0	2.0	0.8	1.0

Notes:
(a) The figures given above are the maximum spacings.
(b) Conduit boxes supporting luminaires or electrical accessories will require separate fixing.
(c) Plastic conduits will require closer spacing of fixings in areas of high ambient temperature and note should be taken of the manufacturer's declared maximum temperature.
(d) The spacings in the table allow for maximum fill of cables in compliance with Appendix A and thermal limits to the relevant British Standards. They assume that the conduit is not exposed to other mechanical stress.
(e) A flexible conduit should be of such length that it does not need to be supported in its run.

▼ **Table G5** Spacings of supports for cable trunking

| Cross-sectional area of trunking, A (mm²) | Maximum distance between supports (m) | | | |
| | Metal | | Insulating | |
	Horizontal	Vertical	Horizontal	Vertical
300 < A ≤ 700	0.75	1.0	0.5	0.5
700 < A ≤ 1500	1.25	1.5	0.5	0.5
1500 < A ≤ 2500	1.75	2.0	1.25	1.25
2500 < A ≤ 5000	3.0	3.0	1.5	2.0
A > 5000	3.0	3.0	1.75	2.0

Notes:
(a) The spacings in the table allow for maximum fill of cables in compliance with Appendix A and thermal limits to the relevant British Standards. They assume that the trunking is not exposed to other mechanical stress.
(b) The above figures do not apply to lighting suspension trunking, or where specially strengthened couplers are used.

(c) Plastic trunking will require closer spacing of fixings in areas of high ambient temperature and note should be taken of the manufacturer's declared maximum temperature.

G5   Conduit bends

522.8.3 The radius of every conduit bend must be such as to allow compliance with the minimum bending radii of cables installed in the conduit (e.g. see Table G2). In addition, the inner radius of the bend should not be less than 2.5 times the outside diameter of the conduit.

G6   Spacing of supports for busbar trunking systems

There is no standard arrangement for the spacing of supports for busbar trunking systems and therefore the manufacturer will provide details of the method of mounting. Supports, where applicable, must be located in specific positions using the spacing dimensions given in the installation instructions.

Appx 8 Different attitudes or mounting arrangements can change the orientation of the conductors within the busbar trunking system and may affect the current-carrying capacity of the system. Where derating is necessary, the manufacturer will provide details of the corresponding mounting factor (k_β) to determine the effective current-carrying capacity of the system using the following formula:

$$I_z = k_\beta \times I_n$$

where:

I_z is the effective current-carrying capacity

k_β is the mounting factor

I_n is the declared rated current.

Appendix H

Maximum demand and diversity

311.1 This appendix gives some information on the determination of the maximum demand for an installation and includes the current demand to be assumed for commonly used equipment. It also includes some notes on the application of allowances for diversity.

The information and values given in this appendix are intended only for guidance because it is impossible to specify the appropriate allowances for diversity for every type of installation and such allowances call for special knowledge and experience. The figures given in Table H2, therefore, may be increased or decreased as decided by the competent person responsible for the design of the installation concerned. For blocks of residential dwellings, large hotels, industrial and large commercial and office premises, the allowances should be assessed by a competent person.

The current demand of a final circuit is determined by adding the current demands of all points of utilization and equipment in the circuit and, where appropriate, making an allowance for diversity. Typical current demands to be used for this summation are given in Table H1. For the standard circuits using BS 1363 socket-outlets detailed in Appendix C, an allowance for diversity has been taken into account and no further diversity should be applied.

▼ **Table H1** Current demand to be assumed for points of utilization and current-using equipment

Point of utilization or current-using equipment	Current demand to be assumed
Socket-outlets other than 2 A socket-outlets and 13 A socket-outlets (note 1)	Rated current
2 A socket-outlets	At least 0.5 A
Lighting outlet (note 2)	Current equivalent to the connected load, with a minimum of 100 W per lampholder
Electric clock, electric shaver supply unit (complying with BS EN 61558-2-5), shaver socket-outlet (complying with BS 4573), bell transformer, and current-using equipment of a rating not greater than 5 VA	May be neglected
Household cooking appliance	The first 10 A of the rated current plus 30 % of the remainder of the rated current plus 5 A if a socket-outlet is incorporated in the control unit
All other stationary equipment	British Standard rated current, or normal current

Notes:

(a) See Appendix C for the design of standard circuits using socket-outlets to BS 1363-2 and BS EN 60309-2 (BS 4343).

(b) Final circuits for discharge lighting shall be arranged so as to be capable of carrying the total steady current, viz. that of the lamp(s) and any associated controlgear and also their harmonic currents. Where more exact information is not available, the demand in volt-amperes is taken as the rated lamp watts multiplied by not less than 1.8. This multiplier is based upon the assumption that the circuit is corrected to a power factor of not less than 0.85 lagging, and takes into account controlgear losses and harmonic current.

The current demand of a distribution system or distribution circuit supplying a number of final circuits may be assessed by using the allowances for diversity given in Table H2, which are applied to the total current demand of all the equipment supplied by that circuit and not by adding the current demands of the individual final circuits obtained as outlined above. In Table H2 the allowances are expressed either as percentages of the current demand or, where followed by the letters f.l., as percentages of the rated full load current of the current-using equipment. The current demand for any final circuit which is a standard circuit arrangement complying with Appendix C is the rated current of the overcurrent protective device of that circuit.

An alternative method of assessing the current demand of a circuit supplying a number of final circuits is to add the diversified current demands of the individual circuits then apply a further allowance for diversity but with this method the allowances given in Table H2 should not be used, the values to be chosen being the responsibility of the installation designer.

The use of other methods of determining maximum demand is not precluded where specified by a competent person.

After the design currents for all the circuits have been determined, enabling the conductor sizes to be chosen, it is necessary to check that the design complies with the requirements of Part 4 of BS 7671 and that the limitation on voltage drop is met.

▼ **Table H2** Allowances for diversity

	Purpose of final circuit fed from conductors or switchgear to which diversity applies	Type of premises		
		Individual household installations including individual dwellings of a block	Small shops, stores, offices and business premises	Small hotels, boarding houses, guest houses, etc.
1	Lighting	66 % of total current demand	90 % of total current demand	75 % of total current demand
2	Heating and power (but see 3 to 8 below)	100 % of total current demand up to 10 A + 50 % of any current demand in excess of 10 A	100 % f.l. of largest appliance + 75 % f.l. of remaining appliances	100 % f.l. of largest appliance + 80 % f.l. of 2nd largest appliance + 60 % f.l. of remaining appliances
3	Cooking appliances	10 A + 30 % f.l. of connected cooking appliances in excess of 10 A + 5 A if socket-outlet incorporated in control unit	100 % f.l. of largest appliance + 80 % f.l. of 2nd largest appliance + 60 % f.l. of remaining appliances	100 % f.l. of largest appliance + 80 % f.l. of 2nd largest appliance + 60 % f.l. of remaining appliances
4	Motors (other than lift motors, which are subject to special consideration)	Not applicable	100 % f.l. of largest motor + 80 % f.l. of 2nd largest motor + 60 % f.l. of remaining motors	100 % f.l. of largest motor + 50 % f.l. of remaining motors
5	Water-heaters (instantaneous type)*	100 % f.l. of largest appliance + 100 % f.l. of 2nd largest appliance + 25 % f.l. of remaining appliances	100 % f.l. of largest appliance + 100 % f.l. of 2nd largest appliance + 25 % f.l. of remaining appliances	100 % f.l. of largest appliance + 100 % f.l. of 2nd largest appliance + 25 % f.l. of remaining appliances
6	Water-heaters (thermostatically controlled)		No diversity allowable[†]	
7	Floor warming installations		No diversity allowable[†]	
8	Thermal storage space heating installations		No diversity allowable[†]	
9	Standard arrangement of final circuits in accordance with Appendix C	100 % of current demand of largest circuit + 40 % of current demand of every other circuit	100 % of current demand of largest circuit + 50 % of current demand of every other circuit	
10	Socket-outlets other than those included in 9 above and stationary equipment other than those listed above	100 % of current demand of largest point of utilization + 40 % of current demand of every other point of utilization	100 % of current demand of largest point of utilization + 75 % of current demand of every other point of utilization	100 % of current demand of largest point of utilization + 75 % of current demand of every other point in main rooms (dining rooms, etc.) + 40 % of current demand of every other point of utilization

Notes:

* In this context an instantaneous water-heater is deemed to be a water-heater of any loading which heats water only while the tap is turned on and therefore uses electricity intermittently.

† It is important to ensure that the distribution boards, etc., are of sufficient rating to take the total load connected to them without the application of any diversity.

Appendix

I

Permitted protective conductor currents

Protective conductor/touch current measurement is an alternative to the in-service insulation test for use if the insulation resistance test either cannot be carried out or gives suspect test results. The current is measured from live parts to earth for Class I equipment, or from live parts to accessible surfaces of Class II equipment.

The current is to be measured within 5 s after the application of the test voltage and must not exceed the values in Table I1.

For practical purposes the test voltage is the supply voltage.

▼ **Table I1** Measured protective conductor/touch current

Appliance class	Maximum current[1]
Mobile or hand-held Class I equipment	0.75 mA
Class I heating appliances	0.75 mA or 0.75 mA per kW, whichever is the greater, with a maximum of 5 mA
Other Class I equipment	3.5 mA
Class II equipment	0.25 mA
Class III equipment	0.5 mA

Notes:

(a) The values specified are doubled if:
  ▶ the appliance has no control device other than a thermal cut-out, a thermostat without an 'off' position or an energy regulator without an 'off' position;
  ▶ all control devices have an 'off' position with a contact opening of at least 3 mm and disconnection in each pole.

543.7 **(b)** Equipment with a protective conductor current designed to exceed 3.5 mA shall comply with the requirements of Regulation 543.7 of BS 7671.

(c) The nominal test voltage is:
  ▶ 1.06 times rated voltage, or 1.06 times the upper limit of the rated voltage range, for appliances for d.c. only, for single-phase appliances and for three-phase appliances which are also suitable for single-phase supply, if the rated voltage or the upper limit of the rated voltage range does not exceed 250 V;
  ▶ 1.06 times rated line voltage divided by 1.73, or 1.06 times the upper limit of the rated voltage range divided by 1.73 for other three-phase appliances.

Appendix J
Standard symbols, units and graphical symbols for general electrical purposes

The following symbols are extracted mainly from IEC 60617, supplemented by references from other Standards.

J1   General symbols

V	volts
A	amperes
Hz	hertz
W	watts
kW	kilowatts
F	farads
p.u.	per unit
ph	phase
p.f.	power factor
L	line
N	neutral
h	hours
min	minutes
s	seconds
=====	direct current (d.c.)
∿	alternating current (a.c.)
2∿	two-phase alternating current
2N∿	two-phase alternating current with neutral
3∿	three-phase alternating current
3N∿	three-phase alternating current with neutral
IPXX	IP number (see Appendix B)
↯	fault (indication of assumed fault location)

Class II appliance (Equipment in which protection against electric shock does not rely on basic insulation only but in which additional safety precautions such as supplementary insulation are provided, there being no provision for the connection of exposed metalwork of the equipment to a protective conductor and no reliance upon precautions to be taken in the fixed wiring of the installation (see BS 2754).)

 Class III appliance (Equipment in which protection against electric shock relies on supply at SELV and in which voltages higher than those of SELV are not generated (see BS 2754).)

 Safety isolating transformer. Class III equipment must be supplied from a safety isolating transformer to BS EN 61558-2-6. The safety isolating transformer will have this identifying mark upon it.

 Isolating transformer

 Protective earth, general symbol (preferred to ⊥)

In the International System of units (known as SI), there are seven base units, as shown below; other quantities are derived from these (see J1.1).

Quantity	Name of base unit	Unit symbol
Length	metre	m
Mass	kilogram	kg
Time	second	s
Electric current	ampere	A
Thermodynamic temperature	kelvin	K
Amount of substance	mole	mol
Luminous intensity	candela	cd

Multiples and sub-multiples of quantities*

10^{18}	exa	E				10^{-3}	milli	m
10^{15}	peta	P	10^{2}	hecto	h	10^{-6}	micro	µ
10^{12}	tera	T	10^{1}	deca	da	10^{-9}	nano	n
10^{9}	giga	G	10^{-1}	deci	d	10^{-12}	pico	p
10^{6}	mega	M	10^{-2}	centi	c	10^{-15}	femto	f
10^{3}	kilo	k				10^{-18}	atto	a

* Powers in steps of 3 are preferred but some others have common usage (e.g. centimetre, cm; decibel, dB).

J1.1   SI derived units

The units of all physical quantities are derived from the base and supplementary SI units and certain of them have been named. These, together with some common compound units, are given here:

Quantity	Unit name	Unit symbol	SI units
Force	newton	N	kg m/s^2
Energy	joule	J	N m
Power	watt	W	J/s
Pressure, stress	pascal	Pa	N/m^2
Electric potential	volt	V	J/C, W/A
Electric charge, electric flux	coulomb	C	A s
Magnetic flux	weber	Wb	V s
Magnetic flux density	tesla	T	Wb/m^2
Resistance	ohm	Ω	V/A
Conductance	siemens	S	A/V
Capacitance	farad	F	C/V
Inductance	henry	H	Wb/A
Celsius temperature	degree Celsius	°C	K
Frequency	hertz	Hz	s^{-1}
Luminous flux	lumen	lm	cd sr
Illuminance	lux	lx	lm/m^2
Activity (radiation)	becquerel	Bq	s^{-1}
Absorbed dose	gray	Gy	J/kg
Dose equivalent	sievert	Sv	J/kg
Mass density	kilogram per cubic metre	kg/m^3	
Torque	newton metre	N m	
Electric field strength	volt per meter	V/m	
Magnetic field strength	ampere per metre	A/m	
Thermal conductivity	watt per metre kelvin	W m^{-1}K^{-1}	
Luminance	candela per square metre	cd/m^2	

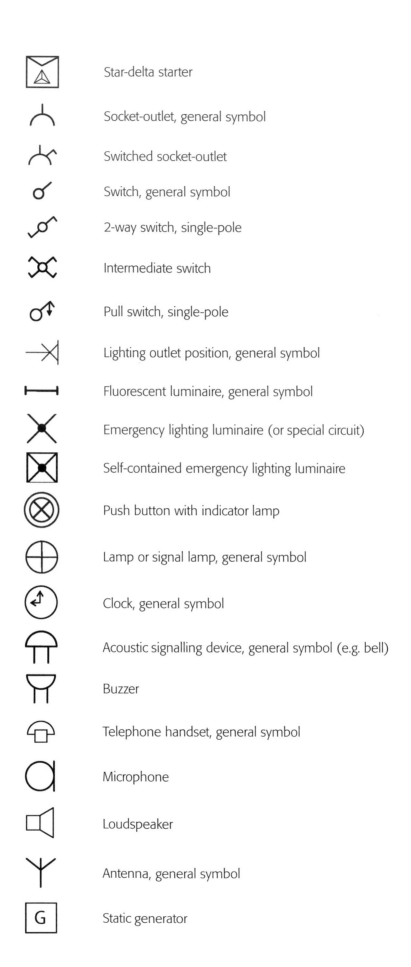

Star-delta starter

Socket-outlet, general symbol

Switched socket-outlet

Switch, general symbol

2-way switch, single-pole

Intermediate switch

Pull switch, single-pole

Lighting outlet position, general symbol

Fluorescent luminaire, general symbol

Emergency lighting luminaire (or special circuit)

Self-contained emergency lighting luminaire

Push button with indicator lamp

Lamp or signal lamp, general symbol

Clock, general symbol

Acoustic signalling device, general symbol (e.g. bell)

Buzzer

Telephone handset, general symbol

Microphone

Loudspeaker

Antenna, general symbol

Static generator

Appendix K

Adresses of associated bodies and industry contacts

Association of Consulting Engineers (ACE)
Alliance House
12 Caxton Street
London SW1H 0QL

Tel: 020 7222 6557

Fax: 020 7222 0750

Web: www.acenet.co.uk

Association of Manufacturers of Domestic Appliances (AMDEA)
Rapier House
40–46 Lamb's Conduit Street
London WC1N 3NW

Tel: 020 7405 0666

Fax: 020 7405 6609

Web: www.amdea.org.uk

ASTA / Intertek

Tel: 01788 578435

Web: www.astabeab.co.uk / www.intertek.com

British Approvals Service for Cables (BASEC)
23 Presley Way
Crownhill, Milton Keynes
Buckinghamshire MK8 0ES

Tel: 01908 267300

Fax: 01908 267255

Web: www.basec.org.uk

British Cables Association (BCA)

Web: www.bcauk.org

British Electrotechnical & Allied Manufacturers' Association (BEAMA)
Westminster Tower
3 Albert Embankment
London SE1 7SL

Tel: 020 7793 3000

Web: www.beama.org.uk

British Electrotechnical Approvals Board (BEAB)

(see ASTA/Intertek above)

British Standards Institution (BSI)

389 Chiswick High Road
London W4 4AL

Tel: 020 8996 9001

Web: www.bsi-global.com

Building Services Research and Information Association (BSRIA)

Old Bracknell Lane West
Bracknell
Berkshire RG12 7AH

Tel: 01344 465600

Web: www.bsria.co.uk

Chartered Institution of Building Services Engineers (CIBSE)

Delta House
222 Balham High Road
London SW12 9BS

Tel: 020 8675 5211

Fax: 020 8675 5449

Web: www.cibse.org

City and Guilds (C&G)

1 Giltspur Street
London EC1A 9DD

Tel: 020 7294 2800

Web: www.city-and-guilds.co.uk

Copper Development Association

5 Grovelands Business Centre
Boundary Way
Hemel Hempstead
Hertfordshire HP2 7TE

Tel: 01422 275705

Fax: 01442 275716

Web: www.copperalliance.org.uk

Department for Business, Innovation and Skills (BIS)

Web: www.bis.gov.uk

Electrical Contractors' Association (ECA)

ESCA House
34 Palace Court
London W2 4HY

Tel: 020 7313 4800

Fax: 020 7221 7344

Web: www.eca.co.uk

Electrical Safety First

Unit 331 Great Guildford Business Square
30 Great Guildford Street
London SE1 0HS

Tel: 020 3463 5100

Fax: 020 3463 5139

Web: www.electricalsafetyfirst.org.uk

Energy Networks Association

6th Floor, Dean Bradley House
52 Horseferry Road
London SW1P 2A

Tel: 020 7706 5100

Web: www.energynetworks.org

Engineering Equipment & Material Users' Association

63 Mark Lane
London EC3R 7NQ

Tel: 020 7488 0801

Fax: 020 7488 3499

Web: www.eemua.org

ERA Technology Ltd

Cleeve Road
Leatherhead
Surrey KT22 7SA

Tel: 01372 36 7345

Fax: 01372 36 7359

Web: www.era.co.uk

Fibreoptic Industry Association

The Manor House
Buntingford
Hertfordshire SG9 9AB

Web: www.fia-online.co.uk

GAMBICA Association Ltd
(Association for the Instrumentation, Control and Automation Industry in the UK)
Broadwall House
21 Broadwall Street
London SE1 9PL

Tel: 020 7642 8080

Web: www.gambica.org.uk

Health and Safety Executive (HSE)

Web: www.hse.gov.uk

Institution of Engineering and Technology (IET)
Michael Faraday House
Stevenage
Hertfordshire SG1 2AY

Tel: 01438 313 311

Web: www.theiet.org

Joint Industry Board for the Electrical Contracting Industry
PO Box 127
Swanley
Kent BR8 9BH

Tel: 03333 218230

Web: www.jib.org.uk

Lighting Association Ltd
Stafford Park 7
Telford
Shropshire TF3 3BQ

Tel: 01952 290905

Fax: 01952 290906

Web: www.lighting.co.uk

Lighting Industry Federation Ltd (LIF)
Ground Floor
Westminster Tower
3 Albert Embankment
London SE1 7SL

Tel: 020 7793 3020

Web: www.lif.co.uk

London Building Control

Web: www.londonbuildingcontrol.org.uk

National Joint Utilities Group (NJUG)

1 Castle Lane
London SW1E 6DR

Tel: 020 3379 3315

Web: www.njug.org.uk

NICEIC Group Ltd

Warwick House
Houghton Hall Park
Houghton Regis, Dunstable
Bedfordshire LU5 5ZX

Tel: 0870 013 0391

Fax: 01582 53 090

Web: www.niceic.org.uk

Royal Institute of British Architects (RIBA)

66 Portland Place
London W1B 1AD

Tel: 020 7307 3700

Web: www.ribafind.org

Safety Assessment Federation

Unit 4, First Floor
70 South Lambeth Road
Vauxhall
London SW8 1RL

Tel: 020 7582 3208

Web: www.safed.co.uk

SELECT (Electrical Contractors' Association of Scotland)

The Walled Garden
Bush Estate
Midlothian
Scotland EH26 0SB

Tel: 0131 445 5577

Web: www.select.org.uk

United Kingdom Accreditation Service (UKAS)

21–47 High Street
Feltham
Middlesex TW13 4UN

Tel: 020 8917 8400

Web: www.ukas.com

IET Standards

Industry-leading standards

IET Standards works with Industry-leading bodies and experts to publish a range of codes of practice and guidance materials for professional engineers, using its expertise to achieve consensus on best practice in both emerging and established technology fields.

See the full range of IET Standards titles at:

www.theiet.org/standards

Centre of Excellence

The Institution of Engineering and Technology

IET Centres of Excellence

The IET recognises training providers who consistently achieve high standards of training delivery for electrical installers and contractors on a range of courses at craft and technician levels.

Using an IET Centre of Excellence to meet your training needs provides you with:

- Courses that have a rigorous external QA process to ensure the best quality training
- Courses that underpin the expertise required of the IET Electrical Regulations publications
- Training by competent and professional trainers approved by industry experts at the IET

See the current list of IET Centres of Excellence in your area at:

www.theiet.org/excellence

The Institution of
Engineering and Technology

IET Membership

By joining a world-leading professional institution you will be
with nearly 160,000 members in 127 countries sharing a
common vision of 'working to engineer a better world'.

Our mission is to inspire, inform and influence by providing a
rich portfolio of products and services designed to meet the
needs of our members at every life:career stage.

As a member you will have access to:

- A 35% discount on BS 7671 (IET Wiring Regulations)
  and associated guidance publications
- Professional registration as EngTech, IEng, CEng, ICT *Tech*,
  CITP or CMgr
- Professional development courses such as specification
  writing, bidding and tendering
- Digital library consisting of thousands of online e-books,
  e-journals and magazines
- E&T magazine
- IET Events
- MyCommunity, the IET's professional online network
- IET Connect, the IET's benevolent fund

Join online today:

www.**theiet**.org/membership

or contact our membership service centre on
+44 (0) 1438 765678